Mahmoud A. ElSohly (Ed.)
Cannabis Chemistry and Biology

Also of interest

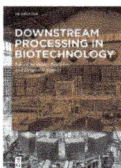

Downstream Processing in Biotechnology
Venko N. Beschkov and Dragomir Yankov (Eds.), 2021
ISBN 978-3-11-057395-4, e-ISBN 978-3-11-057411-1

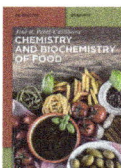

Chemistry and Biochemistry of Food
Jose R. Perez-Castineira, 2020
ISBN 978-3-11-059547-5, e-ISBN 978-3-11-059548-2

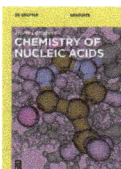

Chemistry of Nucleic Acids
Harri Lönnberg, 2020
ISBN 978-3-11-060927-1, e-ISBN 978-3-11-060929-5

Industrial Biotechnology
Mark Anthony Benvenuto, 2019
ISBN 978-3-11-053639-3, e-ISBN 978-3-11-053662-1

Cannabis Chemistry and Biology

Fundamentals

Edited by
Mahmoud A. ElSohly

DE GRUYTER

Editor
Prof. Mahmoud A. ElSohly
The University of Mississippi
135 Coy Waller Lab Complex
806 Hathorn Road, P.O. Box 1848
University, MS 38677-1848
USA
melsohly@olemiss.edu

ISBN 978-3-11-071835-5
e-ISBN (PDF) 978-3-11-071836-2
e-ISBN (EPUB) 978-3-11-071852-2

Library of Congress Control Number: 2023930117

Bibliographic information published by the Deutsche Nationalbibliothek
The Deutsche Nationalbibliothek lists this publication in the Deutsche Nationalbibliografie;
detailed bibliographic data are available on the internet at http://dnb.dnb.de.

© 2023 Walter de Gruyter GmbH, Berlin/Boston
Cover image: Mahmoud A. ElSohly
Typesetting: Integra Software Services Pvt. Ltd.
Printing and binding: CPI books GmbH, Leck

www.degruyter.com

Contents

List of contributors

Jeffrey S. Block, M.D.
Founder, Nurturing Nature Group Consultants
7299 SW 79th Court
Miami, FL 33143
e-mail: DocBlock@NurturingNature.com
Department of Anesthesiology
University of Miami Miller School of Medicine
1611 NW 12th Avenue (C-301)
Miami, FL 33136
USA
e-mail: DocBlock@Miami.edu

Hemant Lata
National Center for Natural Products Research
School of Pharmacy
University of Mississippi
MS 38677
USA

Suman Chandra
National Center for Natural Products Research
School of Pharmacy
University of Mississippi
MS 38677
USA

Mahmoud A. ElSohly
National Center for Natural Products Research
and
Department of Pharmaceutics and Drug Delivery
School of Pharmacy
University of Mississippi
MS 38677
USA
e-mail: melsohly@olemiss.edu

Shahbaz W. Gul
ElSohly Laboratories Inc.
5 Industrial Park Drive
Oxford, MS 38655, USA
and
Sally McDonnell Barksdale Honors College
and

School of Pharmacy
University of Mississippi
MS 38677
USA

Waseem Gul
ElSohly Laboratories Inc.
5 Industrial Park Drive
Oxford, MS 38655, USA

Mohamed M. Radwan
National Center for Natural Products Research
School of Pharmacy
University of Mississippi
MS 38677
USA

Amira S. Wanas
National Center for Natural Products Research
School of Pharmacy
University of Mississippi
MS 38677
USA

Chandrani G. Majumdar
National Center for Natural Products Research
School of Pharmacy
University of Mississippi
MS 38677
USA

Elsayed A. Ibrahim
National Center for Natural Products Research
University of Mississippi
MS 38677
USA
and
Department of Pharmaceutical Analytical
Chemistry
Faculty of Pharmacy, Suez Canal University
Ismailia 41522
Egypt

https://doi.org/10.1515/9783110718362-203

Jeffrey S. Block

1 The history of cannabis: a story of coevolution and human discovery

Abstract: This chapter covers the known history of Cannabis sativa and its human uses for food, fiber, insight, and medicine, including its role in discovering a key receptor system in animals. The chapter explores the archaeological and written records of where and how the plant evolved and spread with human migration across the globe, from its origins in NE Asia and its travel along the Silk Road to India, Greece and the Middle East, and Europe, before spreading to Britain, the Americas, Africa, and finally Australia. The chapter describes the natural evolution of various phenotypes of cannabis as it adapted to the new environments and how that natural adaptation influenced the diverse uses to which the plant has been put in different cultures, with a primary distinction between fiber hemp and psychoactive cannabis. The challenges of chemical analysis of cannabis lipids are discussed, along with the history of discovery of the endogenous cannabinoid system, a ubiquitous lipid-based receptor system that is responsible for regulating all other bodily systems. Actions of the primary phytocannabinoids, delta-9 tetrahydrocannabinol and cannabidiol, are described, along with the development of modern cannabinoid medications. The change in regulatory approaches in the U.S. are summarized, from colonial requirements to cultivate hemp, to the Marihuana Tax Act of 1937 and prohibition since 1970, to the relegalization of hemp with the 2018 Farm Bill. Effects of modern hybridization and the unintended consequence of prohibition (an increase in THC potency and decrease in CBD in cannabis phenotypes) are also described.

Keywords: Cannabis history, Cannabis sativa, Hemp, Co-evolution, Endogenous cannabinoid system, ECS, Cannabinoids, Cannabinoid receptor, Biochemistry, Lipid chemistry , Anandamide, 2-AG, Tetrahydrocannabinol, Cannabidiol, Cannabis migration, Medicinal cannabis, Medical marijuana, Cannabis policy, Chinese medicine, Ayurvedic medicine

"Nature is the best medicine"[1] is how we translate the Greek phrase attributed to Hippocrates, the classical physician most famous for originating the oath modern doctors

---

1 A more literal rendering of the Greek phrase *"Νόσων φύσεις ιητροί"* is "Nature is the physician of diseases."

Jeffrey S. Block, M.D. Founder, Nurturing Nature Group Consultants 7299 SW 79th Court Miami, FL 33143 e-mail: DocBlock@NurturingNature.com; Department of Anesthesiology University of Miami Miller School of Medicine 1611 NW 12th Avenue (C-301) Miami, FL 33136 USA e-mail: DocBlock@Miami.edu

https://doi.org/10.1515/9783110718362-001

still take. This ancient maxim acknowledges that the human body has powerful natural defenses and mechanisms for healing that can cope with the various maladies that afflict us. The saying also extends to the varieties of plants that have always nurtured and sustained us. One of the most remarkable of those plants that have symbiotically coevolved with humans is *Cannabis sativa*. As controversial and stigmatized as the plant became at the turn of the twentieth century, it is one of the original crops of human cultivation that has provided food, fiber, insight, and medicine for thousands of years. This chapter will explore the history of where and how the plant evolved and spread with human migration across the globe, transforming itself as it adapted to each new climate and was adopted by cultures in every corner of the world. It would stave off starvation, provide clothing and paper, sustain seafaring societies, and become part of the ancient medical wisdom recorded by four ancient cultures' "fathers of medicine": Hippocrates, Chang Chung-ching, Acharya Charak, and Imhotep. In the modern era, it would provide the keys to unlocking the biological secrets of human health, and agricultural ingenuity would again transform the plant, creating chemotypes of cannabis not seen in nature (Figure 1.1).

Figure 1.1: Vitruvian man surrounded by historic cultures' physicians: illustration by Jeannette Aquino.

1.1 Contemporary discoveries meet historical considerations

The genetic diversity of *Cannabis sativa*, due to both natural selection and modern breeding programs, encompasses major differences in chemotypes and phenotypes. Some plants are tall, thin, and reed-like with little intoxicant potential. Others are squat and shrub-like and contain remarkably potent quantities of psychoactive chemicals. The differences have spurred considerable debate about whether cannabis is a single species or several. Some have proposed that three species be recognized: *Cannabis sativa* for the lanky type historically associated with fiber uses, *Cannabis indica* for the shrubby type used for its medicinal and psychotropic properties, and *Cannabis ruderalis* for a short type with little intoxicating potential that developed in Central and Eastern Europe. However, contemporary botanists have largely concluded that all such plants belong to the species *Cannabis sativa* documented by the Swedish taxonomist Carl Linnaeus in his 1753 *Systema Naturea*. Whether a single species or not, the diversity of the plants poses a challenge for distinguishing between types and classifying them accurately and consistently, something that is increasingly important as interest in its medicinal use expands. Several professional groups have established standards of analysis for purity and identity, and the United States Pharmacopeia (USP) formed a Cannabis Expert Panel (which includes the editor of this book) that reached a consensus recommendation to adopt three broad categories of cannabis based on their chemotype of prevailing bioactive chemicals. Those three are varieties in which (1) tetrahydrocannabinols (THC) dominate, (2) cannabidiol (CBD) dominates, and (3) those that contain a more balanced profile of both THC and CBD.

This chapter will consider in detail the history of discovery of those chemicals, known collectively as cannabinoids as well as their significance in human use. It will also recount some of the differences in evolutionary development of these chemotypes and the divergent uses to which they have historically been put. In history, those plant types and uses have been largely bifurcated: one type used primarily for its medicinal and intoxicating properties and another used as a source of fiber, food, and fuel oil. All are properly known as *Cannabis sativa*, a name taken from the Greek, but the names used historically have differed and not just by language (Figure 1.3). In English, "hemp" was the name commonly used for the plant, with "India hemp" sometimes used to distinguish the drug-type from its fiber-type sibling, but many of the medical extracts of cannabis developed by pharmaceutical companies in the late nineteenth century were labeled "hemp tincture." In the U.S. during the 1930s, "marihuana" or "marijuana" was adopted in legislation as the name of the plant, and today U.S. federal law makes a formal distinction between marijuana (the THC-rich chemotype), which is prohibited, and hemp (the low-THC, high-CBD chemotype), which is legal for cultivation and interstate commerce. In this chapter, "hemp" will be used to refer to cannabis plants historically used primarily for fiber or those currently legal for cultivation under U.S. federal law.

Both will receive attention, as will intoxicating cannabis types. Last, understanding the meandering botanical history of cannabis and the promise the plant holds for modern medicine requires us to also consider the much longer history of human biological development, as there lies a remarkable evolutionary coincidence that explains both what we have done to the plant and what it has taught us about ourselves.

Both the use and cultivation of cannabis predates written human history.[2] Today, cannabis grows on all continents except for Antarctica, but it is native to Asia, likely originating in the northeast. From there it traveled with humans south to India, then to the Middle East and Greece, and across Europe with the Romans before being brought to the New World with Spanish colonial explorers and British settlers, and then finally to Australia with the first British settlers there (see Figure 1.2). For many thousands of years, cannabis use and cultivation were centered in Asia, extending from what is now southeastern Russia into India. Cannabis went global with European colonial expansion from the sixteenth to eighteenth centuries, owing in part to the critical role cannabis fiber played in the operation of sailing ships.

Archaeological evidence exists for prehistoric humans using cannabis plants in a variety of ways – stalks for fiber, seeds for food, and flowers and seed husks for medicinal and spiritual purposes. The specific uses differed in prevalence by global regions, influenced at least in part by the differences in plant morphology and chemotype resulting from natural genetic adaptation to the environments in which the plants grew. For instance, the Hindu-Kush region of northern India where the most potent drug varieties of cannabis developed through natural selection is noted for its extreme altitude and cold temperatures, containing most of the world's 8,000 meter peaks and base elevations that exceed the tallest mountains on other continents. The cannabis that developed in that region had considerably higher concentrations of oil-filled trichomes – the tiny bulbous hair-like structures that cover the flowers and leaves of the cannabis plant – which may function as a protective mechanism against the intense ultraviolet light radiation experienced at high altitudes and are much shorter and bushier in shape than the cannabis found in more equatorial climates. Natural selection in those warmer, more humid environments favored tall, lanky cannabis plants that are more reed-like, have less of the intoxicating chemicals, and produce long bast fibers that are exceptionally strong and durable, making them a superior material for making rope, cloth, and paper.

Most of the prehistoric evidence of cannabis fiber comes from sites in China and Taiwan, with one artifact in Taiwan dating to 10,000 BCE. Retting is an ancient process in which the long stems of the plants are soaked in water for a prolonged length of time used to free the bast fibers. The impressions of cannabis seeds offer more dispositive

2 Many good accounts of the evolution and migration of cannabis and histories of its discoveries have been published, including Godwin [1], Abel [2], Robinson [3], Fleming and Clarke [4], Pertwee [5], Russo [6], Iverson [7], and Clarke and Merlin [8].

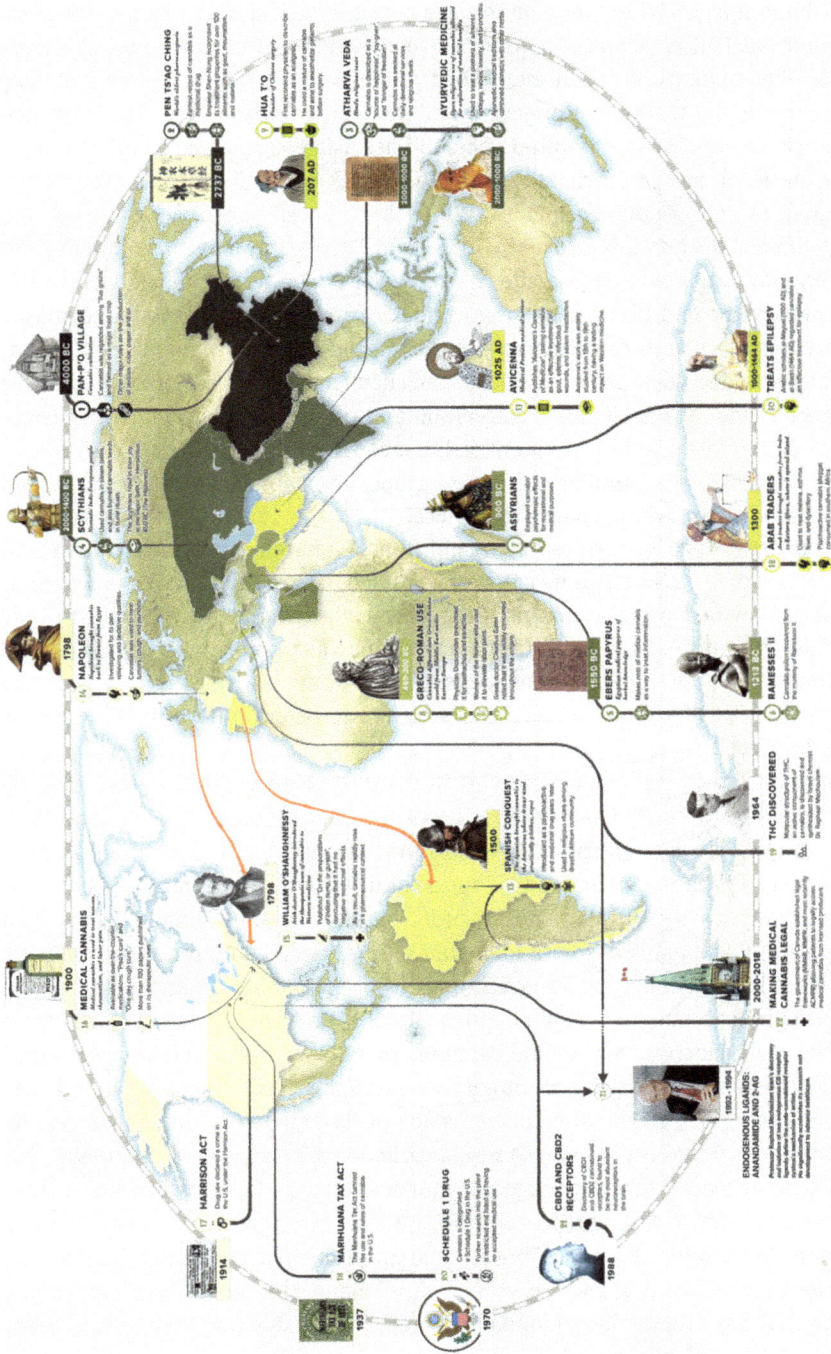

Figure 1.2: Cannabis and human migration around the world.

evidence than impressions of fabric or cordage because the shape and size of the seeds are so distinctive. Retting ponds have provided solid evidence of fiber processing, but the identity of fiber remains has been established primarily by the contexts in which they have been found and the known prevalence of cannabis cultivation and use in ancient China. Fabric impressions on Neolithic pots in East Asia are considered to have been cannabis material, as the fabric is too coarse for silk and the artifacts predate the development of cotton and other cloth. While there is less subsequent written and cultural evidence of cannabis fabric use in ancient Europe, impressions of plant fiber cordage on clay fragments from a site in the Czech Republic that dates back nearly 27,000 years are thought likely to be of woven wild cannabis or nettles, a close cousin. Laboratory analysis can distinguish cannabis fiber from other plant fibers such as flax, but working with degraded fibers is difficult, and such analysis has been rarely used.

Far better evidence of cannabis cultivation and use is found in the form of seeds, which are distinctive and readily identifiable, even when partial or split in halves. Seed remains, which are found primarily in Europe, often appear in batches of partial seeds, indicating areas where retting took place.

Cannabis seeds are a nutritious source of fiber, B vitamins, polyunsaturated fats, and complex proteins. They can be eaten raw, hulled, or ground to make powders or milks. There is evidence that cannabis seeds were collected and stored as emergency rations against famine. In ancient China, cannabis was one of the five staple food crops known as *wu gu*, and stores of seeds have been found there in tombs dating back more than 6,600 years. Cannabis seeds have also been found in Russian tombs from the Iron Age.

Evidence of medicinal use of cannabis in the written record is found in the Chinese herbal guide *Pen-ts'ao* from the first or second century, but archaeological findings show its use for such purposes precedes that. Archaeologists identified a burial site in northern China that contained a Caucasian man who had been buried between 2,400 and 2,800 years ago with a "shroud" of 13 complete female cannabis plants. As a dioecious species, cannabis has both male and female plants, with females producing large flowers rich in therapeutic and intoxicating cannabinoids and other active chemicals. Other archaeological sites along the Silk Road in China's Turpan Basin have uncovered evidence that female cannabis plants were gathered and processed for uses as medicines and spiritual intoxicants. Cannabis seeds were also found buried with a woman in Siberia who showed signs of having died of breast cancer. Its psychoactive use in that region appears to have been rare though not unknown. This contrasts with at least a thousand years of intoxicant use in India, which has a story of the Hindu god Shiva falling asleep in the shade of a cannabis plant, eating some leaves upon awakening, and becoming so enamored of them that cannabis became his favorite food. Shiva is known as the Lord of Bhang, the potent cannabis edible concoction that has been added to food and beverages in India for thousands of years and figures in Hindu festivals and rituals.

1.2 Six millennia of cannabis cultivation

So long as it has rich soil, cannabis needs less human intervention to thrive in concentrated areas than many crops, so it was almost certainly used for thousands of years before it was domesticated. Commerce in cannabis was underway in China about 6,500 years ago, but the consensus view puts the start of cannabis cultivation at roughly 6,000 years ago in several areas of northeast Asia. Areas of northern China have evidence of continuous use of cannabis from the Neolithic to the present. During the Tang Dynasty (618–907 CE), cannabis was a primary crop used for food, fiber, medicine, and spiritual purposes (Figure 1.4). The hemp varieties of cannabis have more cellulose content than wood pulp and resists rot better than cotton. Surviving hemp cloth has been found in Chinese graves from the early eighth century. Papermaking originated in China, and hemp fiber was often a component of ancient paper. Hemp papermaking dates to the Western Han Dynasty in China, a time when the physician Chang Chung-ching (Zhang Zhongjing) created the herbal manual that has influenced traditional Chinese medicine to the present day. The oldest known paper fragment was made from hemp and was discovered in a tomb in China that has been dated to between 140 and 87 BCE. Papermaking reached Europe via the Middle East in the thirteenth century, and hemp remained a mainstay of paper production until the nineteenth century.

Cannabis cultivation appears to have reached India about 3,000 years ago. Unlike China and Asiatic Russia where many cannabis-laden burial areas have been found, archaeological sites in India have revealed no older evidence of cannabis use. As long-distance trade developed along what is known as the Silk Route for one of the primary commodities that made its way from China through India into Europe, cannabis moved with it. Cannabis was cultivated in Greece and the Middle East more than 2,000 years ago. Cannabis rope and fabric-making were widespread throughout Hellenic Greece, and the Greek historian Herodotus mentions its use as an intoxicant or medicine (ca. 100 CE). From the Middle East and Greece, the plant spread to Italy[3] and then north and west to the British Isles in the hands of Vikings, who cultivated it throughout Scandinavia.

Archaeological finds of hemp fabric in Britain are as old as 750 BCE, but evidence suggests that widespread cultivation did not begin there until the incursion of the Anglo-Saxons in the fifth century following the withdrawal of the Romans. Britain's many ancient retting sites for processing hemp fiber show regional concentrations, with pollen records suggesting that hemp comprised 15% of all the acreage under cultivation near the southeastern coast of England.[4]

---

3 The first Roman written record of cannabis use dates to about 100 BCE.
4 Historians estimate that the British fleet of 1588 that defeated the Spanish Armada used 10,000 acres of cultivated hemp. A single ship would require about half a square mile of hemp every two years.

Cannabis was brought to the Americas in the sixteenth century as part of the Columbian exchange – the massive transfer of plants, animals, human populations, and diseases between the New World, the Old World, and West Africa – that occurred in the wake of Christopher Columbus's journey across the Atlantic. Enslaved peoples from Angola brought cannabis with them to Brazil in that era and planted it in the Amazon, where it became known as the "opium of the poor" [9] because it was a less expensive analgesic alternative. In North America, hemp seeds came with the pilgrims on the mayflower. Indeed, British sailing vessels were not only outfitted with acres of sails, rigging, and rope made from cannabis, they also carried hemp seeds with them so crews could plant this essential strategic crop wherever they landed. Hemp was cultivated at the behest of the British navy by settlers from Canada to the southernmost colonies, and many colonies required farmers to reserve a portion of their fields for hemp cultivation, as per a 1632 edict of the Virginia Assembly. By the American Revolution, Virginia and Maryland were the colonies producing the most hemp, and many of the "founding fathers" grew it, including the Virginians Thomas Jefferson and George Washington. While the emphasis of cultivation in North America in that era was for fiber, the medicinal and intoxicating qualities of cannabis were not unknown to the colonists or others in the west. Letters from George Washington to his friend William Pearce refer frequently to Washington's "East India Hemp" plantings at Mt. Vernon, and Washington recommended that his friend plant it too and take care to separate the female plants from the males – a practice aimed at increasing the potency of the females by preventing their fertilization. Fiber cultivation in the U.S. reached its apex in the mid-nineteenth century, with surges during both World War I and World War II to accommodate naval demand. The adaptability of the plant meant that it could be cultivated coast to coast in all the climate zones of the U.S., but the states with the most production historically were Illinois, Indiana, Iowa, Kentucky, Minnesota, and Wisconsin. Commercial cannabis fiber cultivation largely ceased in the U.S. after the Marihuana Tax Act of 1937 rendered it prohibitively expensive, but the prevalence of historical planting meant feral cannabis, often called "ditch weed," was found throughout rural areas for decades.

When cannabis came to Africa is uncertain. Physical evidence of its use dates to the sixteenth century, though some scholars suggest that it reached the continent as much as 4,000 years ago. Traces of cannabis have been found in pipes in Morocco and Ethiopia from the 1500s and similar devices were in use in southwest Africa at that time. Travel accounts and anthropological reports from the seventeenth and eighteenth centuries show that there was a culture of cannabis use in Sub-Saharan Africa then, and by the 1800s, cannabis was a popular intoxicant and used as currency for trade in many areas of the continent, with the apparent exception of West Africa.

As they were in North America, British settlers were responsible for introducing cannabis to Australia. A botanist on the *HMS Endeavour*'s 1770 exploration of Australia, Sir Joseph Banks, returned to England with plans for large-scale hemp cultivation in Australia. The British loss of the war with the American colonists in 1783

Sumerian: A.ZAL.LA

Akkadian: *azallû*

Hieroglyphic: *shemshemet*

Chinese *kanji: ma*

Sanskrit: *bhang*

Persian: *shadanaj*

Hebrew: *kaneh bosem*

Greek: *cannabis*

Figure 1.3: Cannabis written word used in ancient languages.

meant that they had just been cut off from a primary source of cannabis fiber and needed a new one. In 1788, 11 British ships transported the first convicts to what is now Sydney to establish a penal colony and settlement. At that time, the British navy was in dire need of new supplies of rigging materials, so the cultivation of hemp for that purpose was one of the priorities of the settlement in New South Wales.

Cannabis rope and sails were an important strategic resource for colonial expansion. Prior to the advent of nylon ropes, ships at sea relied on hemp for its rigging. The sailing ships of Spain, England, Portugal, and other nations were all rigged with materials made from hemp. The strategic need for large supplies of hemp fiber to equip their navies and trade ships spurred the spread of cannabis cultivation in the New World and eventually Australia. In the nineteenth and twentieth centuries, wars repeatedly cut off supplies of hemp fibers from China and Russia to other nations that needed it for naval rigging. The British introduction of cannabis to Australia and Canada was directed at developing consistent supplies of cannabis fiber. In the United States, attempts to eradicate cannabis drug cultivation in the late 1930s and early 1940s swept up cannabis varietals cultivated for fiber. When World War II started, the U.S. Navy found itself in need of more hemp rope than it could get, in part because demand increased but also because supplies from northern Russia and China were cut off. To supply materials for the war effort, the U.S. Department of

Agriculture prepared an educational video called *Hemp for Victory* to encourage U.S. farmers to again grow fiber varieties of cannabis. What is most remarkable about this educational film is that, less than a decade earlier, Congress had taken a significant step to eliminate all cannabis cultivation, including nonpsychoactive hemp, with the passage of the 1937 Marihuana Tax Act, which levied a prohibitive tax on the plant. It had been successful in not just stopping the cultivation of intoxicating plant varieties that included the THC-rich plants that U.S. pharmaceutical companies had used to manufacture cannabis medicines since the mid-nineteenth century, but also the domestic production of hemp fiber so familiar to colonial North Americans.

Hemp cultivation has historically been centered in Siberia, China, and Canada, but there has been a recent resurgence of hemp planting in parts of the U.S. and Europe (Romania). In the U.S., federal lawmakers have taken steps to reestablish a market in varieties of cannabis with low levels of THC defined in U.S. law as hemp if its dry weight contains 0.3% or less THC. In 2014, the annual Agricultural Act, which covers farm subsidies and regulations and is known as the Farm Bill, established a pilot program for hemp cultivation under the supervision of state departments of agriculture and research universities. The success of that program led to these low-THC varieties of cannabis being classified in the 2018 Farm Bill as an agricultural commodity that is not subject to the Controlled Substances Act of 1970, the omnibus drug bill that established a scheduling system for drugs both licit and illicit that outlawed cannabis in the U.S. The 2018 Farm bill allows each state to submit a production plan to the U.S. Department of Agriculture (USDA) for the licensure and supervision of commercial cultivation of hemp. Less than two years later, the U.S. had at least 157,000 acres under cultivation. As of May 2022, the USDA had approved hemp cultivation plans in 44 states, Puerto Rico, and the U.S. Virgin Islands and directly licensed cultivators in the remaining six states. Native American tribes are also entitled under the Farm Bill to submit hemp cultivation plans, and nearly 60 have received USDA approval at this time. Even though the more potent cannabis rich in THC that is used for medicinal purposes and intoxication remains illegal under federal law, the re-legalization of hemp varieties of cannabis marks a pivot from the prohibitionist impulse of the early twentieth century that drove the passage of the first federal ban in 1937.[5]

---

5 The 2014 and 2018 Farm Bills were not the only legislative attempts at reforming or repealing cannabis prohibition, just the first successful ones. For a history of the bipartisan bills introduced in every Congress since the CSA was enacted in 1970, see Chapter 3 of The Medicalization of Marijuana [10].

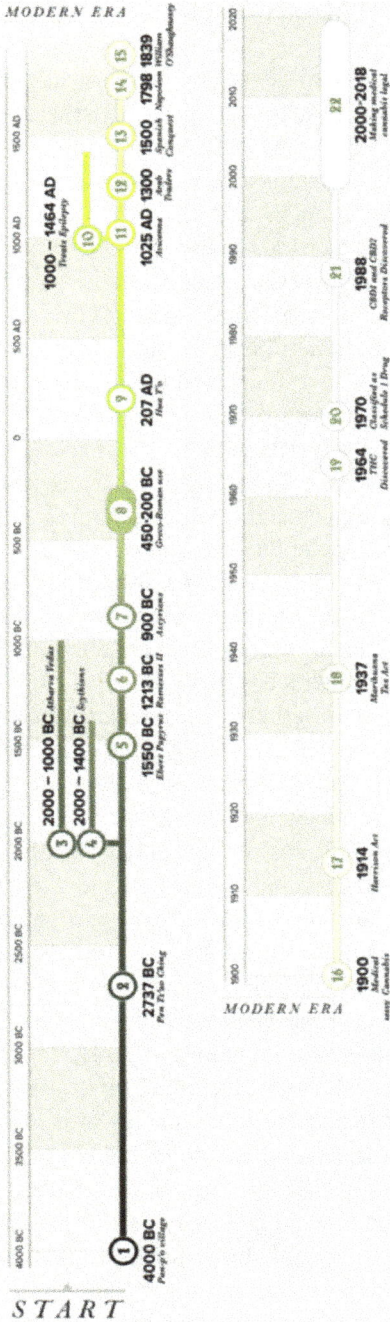

Figure 1.4: Cannabis used by humans: a 6,000-year timeline.

1.3 Documented therapeutic uses

Cannabis preparations are among the oldest medicines, with archaeological evidence of therapeutic use in China as long ago as 2,700 BCE. The earliest written reference to cannabis is found in Assyrian tablets from 700 BCE. The first descriptions of specific therapeutic uses of cannabis are found in the *Pen-ts'ao*, the nearly 2,000-year-old Chinese herbal pharmacopeia attributed in myth to the deity Shen Nung, a text which was supplemented and used by physicians for many centuries. Once the nomadic tribes of northeast Asia carried cannabis across the mountains to India in the Neolithic period, its use expanded beyond the medicinal and ritual, and the edible cannabis preparations known as bhang became integrated into Indian culture. Ancient Indian texts refer to cannabis as one of the "five kingdoms of herbs" and note that it is a remedy for anxiety. The oldest Indian medical text describing cannabis is the *Atharva Veda*, which dates to the same era as the Chinese *Pen-ts'ao*. Cannabis appears several hundred years later in Ayurvedic medical practices, which distinguished between three types of cannabis preparations and therapeutic actions. *Bhang* was prepared from plant leaves, *ganja* from female flowers, and *charas* is the plant's resin, which is notably thick on the surface of cannabis grown at high altitudes. The Indian pharmacopeia describes the various types of cannabis preparations being useful for a variety of conditions and purposes, including analgesia, aphrodisia, inflammation, severe coughing, and digestive and skin conditions [11].

The first western pharmacopeia, the Herbal of Dioscorides from 60 CE, included indications for the use of cannabis, and subsequent herbal manuals in Britain and the U.S. adopted Dioscorides recommendations on cannabis wholesale (as well as a great number of much more suspect ones). Medicinal use of cannabis was known in the West for centuries, as ancient Greek and Roman medical authorities were read by many of the physicians of the seventeenth and eighteenth centuries, but it was the traditional use of cannabis in India that proved indirectly responsible for its popularization in western medicine in the nineteenth century. A British physician in India, William B. O'Shaughnessy, a Victorian-era polymath who was also a professor of chemistry, had observed local cannabis practices with interest and began to conduct systematic investigations. His experiments with administering escalating doses of cannabis to a variety of birds, fish, and mammals, including his patients and students were the first to demonstrate the remarkable safety profile of cannabis [12]. O'Shaughnessy used cannabis to treat a variety of ailments, including cholera, rabies, tetanus, and rheumatism, and he published his findings first in Calcutta and then in England. On his return to London, he brought a large quantity of cannabis with him and engaged a pharmacist in preparing cannabis medicines. His publications and work with drug development influenced a generation of Victorian physicians and launched the commercial development of cannabis tinctures by pharmaceutical companies on both sides of the Atlantic. From 1840 to 1900, western medical journals published nearly 100 articles about medicinal uses of cannabis, most describing its utility as an analgesic and antispasmodic. Most major pharmaceutical companies of the time manufactured several formulations of cannabis medicines. The United States Pharmacopoeia recognized cannabis as a medicine in 1852; it would remain there for 90 years until the pharmaceutical cost of producing cannabis medicines became impossible due to an exorbitant tax Congress had imposed in 1937. Cannabis would remain part of the British Pharmacopeia until 1973.

The Marihuana Tax Act of 1937 passed Congress after short hearings that consisted primarily of lawmakers holding up newspaper clippings with lurid headlines about the insanity and violence that ensued from use of this "new" dangerous drug now dubbed "marijuana." The one voice of dissent was that of Dr. William C. Woodward, a physician and attorney who served as the legislative counsel of the American Medical Association. Dr. Woodward warned that the action Congress was proposing "loses sight of the fact that future investigation may show that there are substantial medical uses for cannabis" [13]. He could not have known that the prohibition of cannabis would delay that research for more than half a century and continue to impede clinical trials and drug development 85 years later. But in his assessment, he was already aware of what future research could find. Contemporary research on cannabis and how it works is revolutionizing our understanding of how our bodies function. What cannabis reveals about the biology of humans and other animals may prove to be one of the most important parts of the plant's history.

1.4 Challenges to cannabis chemistry

Chemists of the nineteenth century had considerable success working with plants such as coca and opium poppies because their active chemicals are water-soluble alkaloids. These plant alkaloids create crystalline solids when combined with acids, making them easily separable from plant material. This led to the creation and commercial development of drugs such as morphine, cocaine, quinine, atropine, and aspirin, forming the foundation of the modern pharmaceutical industry. Cannabis is not so cooperative.

Instead of alkaloids, the primary chemicals in cannabis are lipids or fats. The distinctive trichomes that cover the flowers and leaves of the plant are tiny oil-filled globules. Those oils are easily extracted using a variety of techniques, including such traditional methods as hand rubbing the plants or boiling them in water or milk with other oils such as butter and contemporary methods such as agitating plant material in ice baths or separating the oils using butane or CO_2 techniques. Getting the oil off the plant is one thing; persuading it to divulge its makeup is another matter. Because lipids are not water soluble (as every cook knows, oil and water do not mix), combining lipids with an acid in water will not precipitate a solid. Unlocking the chemistry of phytocannabinoids had to wait for the development of new techniques.

The first phytocannabinoid to be isolated was cannabinol (CBN), the chemical to which the psychoactive Δ^9-THC degrades. While that was achieved in the late nineteenth century, the chemical structure was not elucidated. The next was cannabidiol (CBD), which was identified in 1940 in the laboratory of Roger Adams. Adams was also the first to identify the tetrahydrocannabinols in 1942, though he was not able to fully describe the chemical structure or biosynthesis of either THC or CBD. It would be another 20 years before that was achieved in the laboratory of Israeli chemist Raphael Mechoulam. Mechoulam and chemists working with him described first the molecular stereochemistry of CBD in 1963 and then THC the following year. After thousands of years of human use, scientists could now start investigating not only *how* cannabis works but also *why* it impacts human beings.

Searching for answers about biology based on plant evidence has been successful before. The scientific process of identifying plant compounds that produce effects, then using that knowledge to find bodily systems on which they act, had yielded results with plant alkaloids; but that process was rarely quick or easy. After nearly a century of commercial use treating pain, we knew that aspirin had analgesic properties long before scientists identified prostaglandins as the pharmaceutical answer for how and why aspirin works. It took even longer to understand opium's analgesic effects, which were known for thousands of years before endorphins and enkephalins were identified.

What we have learned about the chemistry of *Cannabis sativa* has led to important discoveries about the central biological regulatory mechanisms of all animals, not just people. Following the identification of the primary phytocannabinoids (THC and CBD), the two most prominent active chemicals in the plant, a search ensued for how those chemicals produce their effects in humans and other animals.

Scientists understood that the drugs made from plant alkaloids such as morphine exploit receptor systems in the body, mimicking natural body chemicals (endogenous ligands) just well enough to cause the receptors to fire and produce a discernible effect such as pain reduction or sedation. In the case of cannabis, the psychoactive effects of THC and, to a lesser extent, CBN were the targets of investigation. Where was the endogenous cannabinoid system (ECS) and how does it operate? Rather than work with the phytocannabinoids, which were difficult to isolate, chemists used new synthetic cannabinoids developed by Pfizer to investigate biologic action. Chemists developed two in vitro assays for cannabinoids that could reveal their presence in a laboratory test tube. One assay looked for a chemical, adenylate cyclase, that cannabinoids inhibit. The other used a radioactive material that binds to the cannabinoids. Using these two methods, Allyn Howlett's lab at St. Louis University found conclusive evidence of a G-protein-coupled cannabinoid receptor in the mid-1980s, with William Devane collaborating on the radio-labeled ligand assay. This receptor was called "cannabinoid receptor" in recognition of how the study of Cannabis led to its discovery. By 1990, these in vitro findings were confirmed with two in vivo experiments. Tom Bonner's lab at the National Institute of Health successfully cloned a rat CB1 receptor that year, while Catherine Gérard and colleagues achieved the same in Brussels with a human CB1 receptor. In 1993, Sean Munro's laboratory in Cambridge discovered a second G-protein-coupled cannabinoid receptor, CB2. Now that these two cannabinoid receptors had been identified, work could begin on developing ways of monitoring when and how they are activated or blocked. The search for the native chemicals that those receptors use began.

Mechoulam's lab would claim those discoveries, too. The first endogenous cannabinoid was isolated in 1992 by two chemists in Mechoulam's lab, William Devane and Lumír Hanuš. Identified chemically as arachidonoyl ethanolamide, the research team somewhat whimsically named it anandamide after the Sanskrit word for bliss and joy, *ananda*, a choice which has proven apt for describing the cannabinoid's role in sustaining a balanced, healthy life. Later that year, Mechoulam's lab discovered 2-arachidonoylglycerol, which got the more prosaic abbreviation 2-AG. Anandamide proved to be what is known as a partial agonist at CB1, meaning it does not fully activate the receptor. 2-AG is the more complete agonist at both CB1 and CB2 and appears to be more responsible for ECS function. Of the two, anandamide most resembles the phytocannabinoid THC in its partial receptor agonist impact. This is not to say that the phytocannabinoid and its endogenous counterparts are chemically similar. They are not, as the accompanying figure illustrates. What they do have in common are remarkably similar lipid solubilities and that unique chemistry is what allows the cannabis plant's bioactive constituents to activate cannabinoid receptors (Figures 1.5 and 1.6).

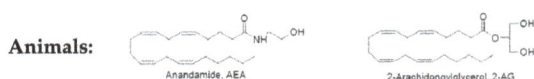

Animals:

Anandamide, AEA   2-Arachidonoylglycerol, 2-AG

When compared to natural *endocannabinoids* (Anandamide & 2-AG), the Cannabis plant's *phytocannabinoids* (THC & CBD) evidence very different molecular structures.

However, these chemicals share remarkably *similar lipid solubilities*, thus explaining their bioactivity on animal Cannabinoid receptors.

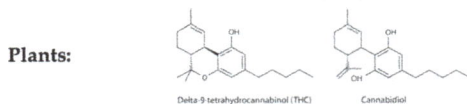

Plants:

Delta-9-tetrahydrocannabinol (THC)   Cannabidiol

Figure 1.5: Comparison of cannabinoid molecular structures.

Just as phytocannabinoids differ from other biologically active plant compounds by being lipids or fat-based, so too do the endogenous cannabinoid receptors differ from other receptors and their endogenous ligands in being the only lipid-based neurotransmitter system. Both anandamide and 2-AG are generated on-demand within a lipid ring that keeps the endocannabinoids and their receptors sequestered from the aqueous environments in which they reside. The ECS also appears to be the only system with both dual receptors and dual ligands, indicating a different degree of modulatory capacity than systems that are effectively on or off. Cannabinoid receptors are involved in nearly all physiological functions of the peripheral and central nervous systems (CNSs) as well as various peripheral organs. Many types of receptors can influence the activity of other receptors by enhancing or inhibiting them, but the ECS operates continuously, fine-tuning the action of all other receptor systems.

Since the discovery of 2-AG and anandamide, several other apparent cannabinoid receptor agonists have been identified. The work of better describing the complex molecular mechanisms for the biosynthesis, activation, and degradation of these lipid mediators is far from complete. Nonetheless, the picture of this critical system is becoming clearer.

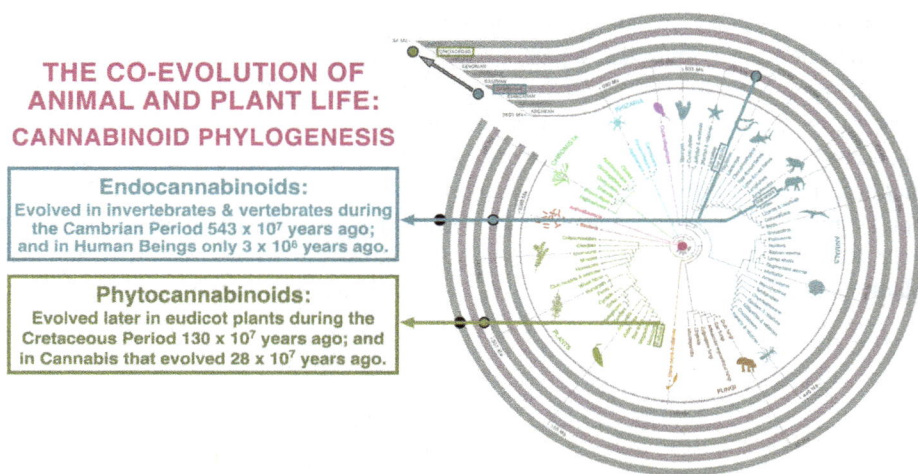

Figure 1.6: Cannabinoids and coevolution.

What we have discovered so far is that all animals from 600-million-year-old sea squirts to humans have this unique receptor system. All animals, vertebrates and even invertebrates, have an ECS. Only insects and arachnids do not. Because it was identified through the investigation of how plant-derived cannabinoids exert their effects, it is known as the ECS. We have also discovered that cannabinoid receptors are not just the most ubiquitous in the body, far out numbering all others, but it is also the master regulatory system, responsible for managing the function of all other receptor systems to maintain homeostasis, the healthy balance point of an organism's various systems. The ECS regulates appetite and metabolism, pain response, mood, memory, immune function, and the sleep-wake cycle of circadian rhythms, among other things. The role of the ECS is so central to how our bodies not just function under normal circumstances but respond to illness and injury that a review by researchers at the National Institute of Health concluded that "modulating ECS activity may have therapeutic potential in almost all diseases affecting humans" [14].

The ECS achieves this remarkable mastery by operating in a completely different fashion than other receptor systems. Rather than signaling "downstream" across a neuron's synapse, the ECS ligands are postsynaptic and travel retrograde, sending signals backward as a feedback loop on neurotransmitters, telling neurons to produce more or less of endogenous chemicals such as serotonin or endorphin. That modulatory action combined with the centrality of endocannabinoids to all the functions of the body helps explain why cannabis preparations are among the oldest medicines. It is also an essential part of the story about how all successful life forms have evolved on Earth, evidencing their abilities to cope with the dangers in their environment.

The ECS in humans and other animals is continuously balancing five general functions that are essential to sustaining life from birth to death. The general purpose of the ECS is to support the ability to eat, sleep, relax, and forget as well as mitigate the consequences of disease as age advances. That our growth and development depend on the right balance of eating and sleeping is obvious, and this is a critical role of CB_1 receptor function. The ability to rest and recover from physical and psychological stress also stems from endocannabinoids and includes the ability to forget. New memories are processed in the brain's hippocampus, an area rich with CB_1 receptors that helps to filter out memories from traumatic experiences – this is a neurocognitive protective mechanism essential for behavioral health. Some physical, psychological, and even spiritual experiences can be intolerable. Forgetting them, or at least remembering them less acutely, is a cannabinoid-based coping mechanism. When it fails, pathological pain-avoidance behavior can result, such as is seen in people suffering from posttraumatic stress disorders (PTSDs). On the flip side, pleasurable experiences that remain too prominent in memory can also produce problems when they lead to an inability to control pleasure-seeking behavior and addiction results. These are roughly what appear to be some of the critical functions of CB1 receptors.

CB2 receptors provide different functions that are no less critical. Instead of being concentrated in the CNS, CB2 receptors are primarily found in the gut where they manage immune function and provide the brain:gut connection that is the subject of much current investigation. Interestingly, CB2 receptors appear on demand in the CNS in response to injury such as a traumatic brain injury or the development of Parkinsonism. CB2 receptors help the body mount defenses and corrective physiologic action. It appears that CB2 receptors are integral to the body's ability to manage diseases, halting progression that would otherwise result in death and rendering those diseases chronic.

Also worth noting is where cannabinoid receptors are *not* found: in the part of the brain responsible for breathing. This is unique from the receptors to which opioids and many other drugs attach and helps explain the unusual safety profile of cannabis and phytocannabinoids. Opioids, alcohol, and the like can kill by overloading brain receptors in ways that stop breathing. Cannabinoids cannot. Both phytocannabinoids and endogenous cannabinoids are responsible for a broad range of receptor activity. Where phytocannabinoids and endogenous cannabinoids differ is the locations they affect and their duration of action. Endogenous cannabinoids are created and degraded rapidly on-site and on-demand. Phytocannabinoids typically produce system-wide effects that persist far longer.

The evolutionary coincidence of cannabis and the much older endogenous system developing independently, yet able to interact, is astonishing. The ECS is a lipid-based receptor system, unlike others. The lipid cannabinoids the plant developed are similarly unusual. The ECSs universal in animals developed at least 600 million years ago, as evidenced by the fossil record of sea squirts which have survived to the present day. Then, after at least 550 million more years of animal evolution, the plant kingdom produces the botanical family that includes cannabis, a plant that can interact with that regulatory system. When exactly cannabis emerged is unclear from the fossil record, but its closest genetic relative, hops, does not appear until around 6 million years ago [15]. That the concatenation of random repeated rolls of the genetic dice could produce a plant with chemicals that dovetail so neatly with animal biology seems like a coevolutionary miracle, even if it took nearly 600 million years and trillions of genetic iterations to achieve.

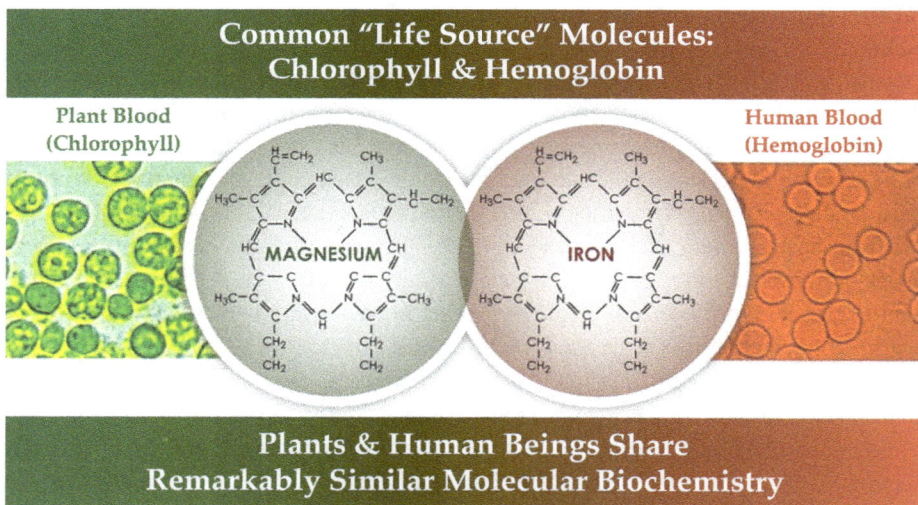

Figure 1.7: Chlorophyll and hemoglobin: life kingdom defining molecules share remarkably similar biochemistry.

The biochemical evolutionary links between plants and animals become apparent when comparing the critical energy molecules of plants and animals: chlorophyll and hemoglobin. Chlorophyll enables plants to process light energy through photosynthesis, while hemoglobin allows animals to derive energy from oxygen exchange. The molecules are noticeably similar in structure, with merely a difference of 2^+ cation. Substitute iron for the magnesium at the core of chlorophyll and you have hemoglobin (Figure 1.7).

That two essential components of life in separate kingdoms could share such similarity of molecular structure suggests that they may be descendants of a common chemical ancestor in the evolutionary lineage.

As old as the ECS is in the animal kingdom and as young as cannabis is, cannabis was already well established in NE Asia by the time Homo sapiens appeared in the hominoid lineage, only 300,000 years ago. (Given that humans may have adopted cannabis use 50,000 years ago, we seem to have learned quickly.) Many plants have developed attributes that repel insect, animal, and disease threats while attracting animals that will disseminate their seeds. Humans have been this special plants' most effective co-evolutionary partner, carrying seeds across oceans and modifying the environment on their behalf.

1.5 Complicated clinical correlations

Δ^9-THC is the phytocannabinoid that has received most of the attention in modern research as it is the primary psychotropic chemical in the plant and the one tied too much of its therapeutic action. In recent years, however, attention has also turned to cannabidiol (CBD) because it can also produce therapeutic effects such as fighting inflammation and has shown potential for treating anxiety and psychosis. Yet the two cannabinoids are dramatically different in ways that shed further light on how the ECS functions. In some respects, CBD and THC are oppositional. While THC is a partial agonist of CB1 receptors, activating them by attaching to their orthosteric sites, CBD is not an agonist at all. Instead, it is an allosteric modulator of CB1 receptors, meaning it attaches to a different site on the receptor and changes its conformation, or shape, in ways that not only reduce the effects of THC but also the endogenous cannabinoid anandamide. CBD also activates complementary receptor functions, inhibits the digestive metabolism of THC edibles, and potentiates the amino acid tryptophan, which is famously responsible for elevated mood and relaxation following a turkey dinner. Together, these properties make CBD a moderator of the psycho-toxic effects of THC's central impact, including counteracting the anxiety and paranoia that can result at higher doses.

At CB2 receptors, CBD acts as a partial agonist, much as THC and anandamide are at CB1, but CB2 receptors are those more prevalent in the gut and immune cells rather than the CNS. CBD also interacts with many other biologic components, affecting their function, including other G-protein-coupled receptors (GPCRs), transient receptor potential (TRP) channels, enzymes, ion channels, transporters, and nuclear hormone receptors. Of these various targets, the TRP Vanilloid One channel (TRPV1) is particularly consequential. Research continues, but TRPV1 activation appears to contribute to the control of inflammation, pain, and temperature sensation. TRPV1 can be stimulated by anandamide, CBD, and other less prevalent phytocannabinoids. Downregulation of

TRPV1, which can result from being activated over time, may contribute to CBD's analgesic and anti-inflammatory effects. Those effects have spurred the development and marketing of a host of topical CBD-infused preparations that people rub on arthritic hands and sore muscles.

The mechanism by which CBD can control seizures remains unknown, but recent research proposes that it may result from its action across a range or targets, including TRPV1, an orphan GPCR, and a nucleoside transporter [16]. Regardless of how it works, CBD has been lauded as a miracle drug for the treatment for severe childhood seizure disorders. CBD extracted from cannabis has been shown in clinical trials and numerous anecdotal reports to dramatically reduce seizures associated with several disorders, even among patients who have responded to no other medications. CBD also has virtually no side effects, in comparison to several significant side effects of conventional anti-seizure medications. As a result, the U.S. Food and Drug Administration (FDA) approved the sublingual CBD medication Epidiolex™ in 2018, making it the first plant-derived, standardized, and dose-controlled medication approved in the United States.

It was not, however, the first standardized, dose-controlled cannabis medication. That distinction belongs to another sublingual extract known as Sativex™, which contains a 1:1 ratio of CBD and THC. The drug has not received FDA clearance as of 2022, but it has been approved for use as an analgesic in the U.K., Canada, and other countries. In addition to these two medicines manufactured directly from cannabis plants, there are two FDA-approved synthetic cannabinoid medicines – dronabinol (Marinol™) and nabilone (Cesamet™). Dronabinol is synthetically manufactured THC and is chemically identical to the phytocannabinoid. It has been approved to treat nausea and vomiting associated with cancer chemotherapy and wasting from appetite loss associated with HIV/AIDS. Nabilone, which is also used in the treatment of nausea and vomiting, is chemically similar to THC but more potent. Both drugs were developed as an outgrowth of what we have learned about the chemistry of the cannabis plant, and future drug development will surely continue to exploit the chemical lessons of the plant. Some may mimic plant chemicals, as these two do, but others are being developed to more selectively target the ECS. This holds great promise for treating virtually all diseases, but it is not without dangers.

The signal example of the risks associated with manipulating the ECS is the anti-obesity drug rimonabant (Acomplia™). Brought to the pharmaceutical market in Europe in 2006, rimonabant suppresses appetite by acting as an inverse agonist at CB1. This drug's ability to occupy CB1 receptors, antagonize its effects, and suppress natural receptor action did, indeed, suppress appetite and produce weight loss in the people who took it. Unfortunately, the ECS functions of CB1 receptors are considerably more far reaching than just appetite. Blockading CB1 meant disabling a portion of the master regulatory system in the body, which not only includes GI function but also sleep cycles and mood regulation. Side effects included nausea and diarrhea, insomnia, anxiety, depression, and suicidal ideation. After several suicides among people taking rimonabant, it was withdrawn from the market in 2008. This cautionary tale highlights the complexity of challenges facing pharmaceutical companies as they seek to capitalize on the insights gained from studying human uses of cannabis.

That long history of therapeutic use points to a way forward. Though research has been limited by the constraints of drug scheduling and other administrative barriers, the evidence in support of cannabinoids as a treatment for a variety of medical conditions is well summarized in a pair of reports commissioned by the U.S. government – one from the Institutes of Medicine in 1999 [17] and the other from the National Academies of Sciences, Engineering, and Medicine in 2017 [18]. These reviews include preclinical laboratory and animal studies as well as experimental and clinical trials with humans. Empirical data are limited or of low quality for the use of cannabinoids in treating many of the conditions for which there are anecdotal reports of efficacy, but there have been a number of high-quality double-blind, placebo-controlled clinical trials showing that cannabis is effective in the treatment of chronic pain. Multiple sclerosis is another condition for which the experts at the National Academies found "substantial evidence" of cannabinoid medicines being efficacious.

As has been noted in the discussion of the plant's evolution, to say cannabis or cannabinoids are effective in treating one thing or another is to beg the question of exactly which variety of cannabis or which ratio of particular cannabinoids is being discussed. And since the effects of cannabinoids, particularly THC, are both highly dose-dependent and often paradoxical – with a low dose producing one effect and a high dose producing the opposite – any consideration of clinical applications or even so-called recreational use must include a careful accounting of exactly how much of any given cannabinoid is being consumed. Even then, the total THC or CBD content of any product does not tell the whole story of its therapeutic effect. Whole-plant products containing many other chemicals acting in concert, including 100-plus minor cannabinoids and the terpenes and flavonoids that give cannabis varieties their distinctive aromas and flavors are at play. Myrcene, for instance, a terpene also found in hops, lemongrass, and basil, is found in high concentrations in *Cannabis indica* varieties known for their hypnotic, sedative effects. Those effects were previously thought to result from the cannabinoid profile, but research has shown that it is the myrcene driving that experience, albeit likely potentiated by the cannabinoids. Terpenes seem to be the chemicals largely responsible for the diverse effects of different cannabis cultivars, though the mechanism of action remains to be discovered.

Despite this new understanding, medical cannabis products are still often labeled and marketed based on the phenotypes of cannabis plants, with squat indicas being identified as sedative and lanky sativas as energizing. As discussed earlier, there is a difference in the evolutionary development of plant phenotypes, with indica cannabis varietals developing in northern latitudes (such as the highlands of India, for which it is named), and sativa types developing in warmer, low-altitude equatorial regions. But those evolutionary distinctions have largely been lost to the human-powered migration of cannabis around the planet and even more so to the sophisticated modern breeding programs that have crossed and back-crossed different phenotypes and chemotypes.

Recent genomic analyses of cannabis plants reveal that almost all commercial cannabis today are hybrids of the two historical types and share similar cannabinoid profiles. What is different between varieties are the terpenes.

The modern process of hybridization has done more than flatten evolutionary differences between the historical landraces identified as *Cannabis sativa* and *Cannabis indica*; it has sharply increased the potency of the intoxicant class of cannabis. Ironically, this is one of the byproducts of the "War on Drugs" that ramped up in the 1980s and 1990s. When most of the cannabis consumed in America was imported from Mexico, the harvested flowers were typically from fertilized female plants, making them seedy and relatively low in THC by current standards. Then zealous enforcement of prohibition stopped much of the cannabis trafficking across the southern border in the 1980s, and American cannabis enthusiasts turned to cultivating it themselves, often in small amounts in tightly controlled indoor operations where culling males was easy and more potent "sinsemilla" became the norm.[6] Their underground gardens (often literally so, in basements) also allowed cultivators to create environments far richer in nutrients, light, and carbon dioxide than nature provides, spurring unnaturally robust plant growth. These enthusiasts also turned their attention to the project of breeding more potent plants, with considerable success. Plant author Michael Pollan has written that he suspects that the greatest gardeners of his generation were those growing underground cannabis. U.S. law enforcement scrutiny expatriated many of those U.S. gardeners to Amsterdam, where cannabis laws had loosened in the mid-1970s. There they found kindred spirits and commercial opportunities for designer cannabis genetics. Dutch breeders and their American and Canadian counterparts created commercial seed banks and mailed the new genetics around the world.

The impact of that project can be seen clearly in the change in average potency of cannabis. An analysis of cannabis seized in the U.S. from 1995 to 2014 found that the percentage of THC tripled, while the ratio of CBD to THC dropped from 1:14 to a far more diluted 1:80 [19]. More THC and less CBD results in cannabis products that express psychoactivity; but as with alcohol, potency does not equal dose. The difference between a liter of beer and a liter of whiskey is clear, and users generally understand how to adjust the quantity consumed to achieve its desired effect. Nonetheless, if only high-potency cannabis products are available, consumer behaviors may be unfavorably impacted.

---

6 Without male plants to fertilize them, female cannabis plants cannot produce seeds (hence the Spanish "without seed" name), but the heroic effort they make to capture any male pollen substantially increases the density of the cannabinoid-filled trichomes, making the flowers considerably more potent. This technique has become so prevalent, that the "sensimilla" distinction used to mark the strongest cannabis has disappeared. All commercial cannabis is without seeds now.

UNBALANCED "MODERN" PLANTS *DEVOLVED* **FROM TRADITIONALLY BALANCED PLANTS**

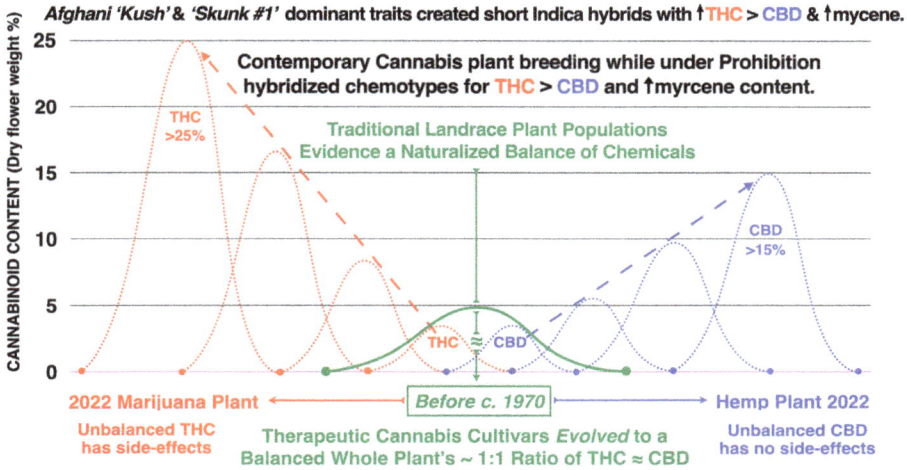

Figure 1.8: A half-century of THC and CBD hybrid potency.

Cannabis bred to maximize intoxication may sacrifice therapeutic potential. Importantly, medicinal benefits from THC are often achieved with relatively low doses, and since effects can be both dose-dependent and paradoxical, more is generally not better. Nevertheless, potential clinical exceptions appear to be for the treatment of patients living with PTSDs or patients dealing with issues associated with cancer pain, both of which appear to benefit from higher doses of THC.

On the other end of cannabis genetics is hemp, which has negligible amounts of THC but now includes varieties bred to contain considerable amounts of CBD. The new U.S. market in hemp includes products derived from it such as CBD, which is receiving increasing attention for its potential in treatment but has fundamentally different mechanisms of action from THC and a different efficacy profile. Between these two ends of the THC–CBD spectrum lie the "intermediate" chemotype identified in the recommendations of the USP Cannabis Expert Panel, those plants with a more balanced THC:CBD ratio closer to 1:1 (please see Figure 1.8). That appears to be the cannabinoid ratio found in clinical trials to be most effective in treating pain. Other findings indicate that substantial CBD amounts mitigate the potential for adverse side effects from THC. There is also evidence that the most effective cannabis products are not isolates of any single cannabinoid but rather a combination of various cannabinoids and terpenes which work in concert to yield a synergistic "entourage" or ensemble effect superior to its individual parts (Figure 1.9).

The complex chemistry of cannabis makes it exponentially more challenging to research than single-molecule botanical medicines, and its widespread impact on the ECS, a system with reach into every aspect of physiology, further complicate matters. Misclassification of cannabis under the U.S. Controlled Substances Act and the UN

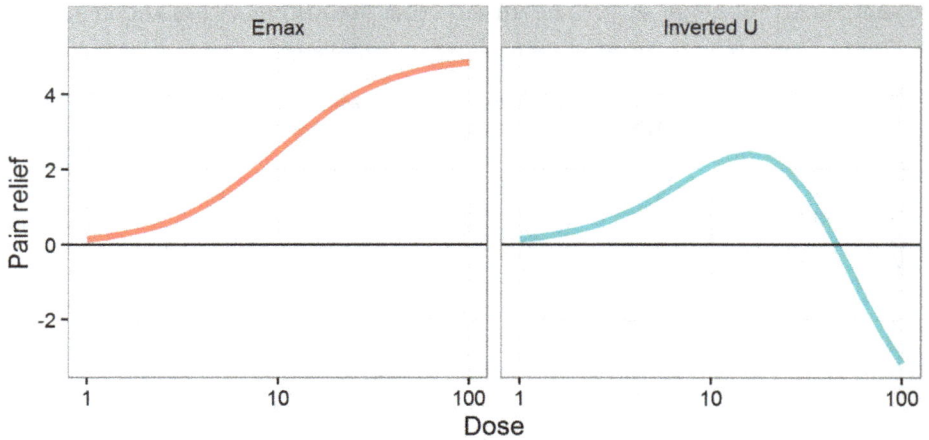

Examples of dose-response relationships. An ideal pain drug would look like the Emax pattern on the left.
Cannabis dose-response for pain relief looks like the pattern on the right (but only for some types of pain).

Figure 1.9: Ideal versus paradoxical (inverted U) dose-response curves.

Single Convention on Narcotics, where it is treated similar to dangerous single-molecule drugs with no medical use, has constrained drug policy around the world and further limited research and new drug development. Compounding the problem is both stigma and ignorance within the medical and wider scientific community. Though the trend to normalization of cannabis use is strong, cannabis still carries negative associations that affect professional standing and medical education. Physicians and other medical professionals will benefit from learning more about cannabis and the ECS, if for no other reason that they get questions from their patients about using cannabis medicinally or taking CBD. This better understanding of the history of cannabis and what it teaches us about ourselves helps strip away the myth and stigma that has obscured the gifts of this extraordinary co-evolutionary relationship.

References

[1]   Godwin H. The ancient cultivation of hemp. Antiquity 1967; 41(161): 42–49.
[2]   Abel EL. Marihuana: The first twelve thousand years. New York, NY, USA, Plenum Press, 1980.
[3]   Robinson R. The great book of hemp: The complete guide to the environmental, commercial, and medicinal uses of the world's most extraordinary plant. Rochester, VT, USA, Park Street Press, 1996.
[4]   Fleming MP, Clarke RC. Physical evidence for the antiquity of *Cannabis sativa* L. Journal of the International Hemp Association 1998; 5(2): 80–95.
[5]   Pertwee RG. Cannabinoid pharmacology: The first 66 years. British Journal of Pharmacology 2006; 147(Suppl 1): S163–S171.

[6]   Russo E. History of cannabis and its preparations in saga, science, and sobriquet. Chemistry & Biodiversity 2007; 4: 1614–1648.
[7]   Iverson LL. The science of marijuana, 2nd ed. New York, NY, USA, Oxford University Press, 2008.
[8]   Clarke R, Merlin M. Cannabis: Evolution and ethnobotany. Univ of California Press, 2016. Berkeley, CA, USA.
[9]   Caceres Guido P, Riva N, Calle G, et al. Medicinal cannabis in Latin America: History, current state of regulation, and the role of the pharmacist in a new clinical experience with cannabidiol oil. The Journal of the American Pharmacists Association 2020; 60(1): 212–215.
[10] Newhart M, Dolphin W. The medicalization of marijuana: Legitimacy, stigma, and the patient experience. New York, NY, USA, Routledge, 2019.
[11] Tavhare S, Acharya R. Bhanga (*Cannabis sativa* L.) as an activity potentiator in Ayurvedic classics and Indian alchemy (Rasashastra): A critical review. International Journal of Ayurvedic Medicine 2016; 7(3): 136–152.
[12] Mills JH. Cannabis Britannica: Empire, trade, and prohibition, 1800–1928. New York, NY, USA, Oxford University Press, 2003.
[13] U.S. Congress, House of Representatives, Committee on Ways and Means, Taxation of Marihuana, Hearing before the Committee on Ways and Means, 75th Cong., 1st sess., May 4, 1937.
[14] Pacher P, Kunos G. Modulating the endocannabinoid system in human health and disease – Successes and failures. The FEBS Journal 2013; 280(9): 1918–1943.
[15] McPartland JM. Cannabis systematics at the levels of family, genus, and species. Cannabis and Cannabinoid Research, Dec 2018. 3(1): 203–212.
[16] Gray RA, Whalley BJ. The proposed mechanisms of action of CBD in epilepsy. Epileptic Disorders 2020; 22: S10–S15.
[17] Joy JE, Watson SJ, Jr., Benson JA, Jr. (Eds.). Marijuana and medicine: Assessing the science base. Washington, DC, USA, The National Academies Press, 1999.
[18] National Academies of Science, Engineering and Medicine. The health effects of cannabis and cannabinoids: The current state of evidence and recommendations for research. Washington, DC, USA, The National Academies Press, 2017.
[19] ElSohly MA, Mehmedic Z, Foster S, Gon C, Chandra S, Church JC. Changes in cannabis potency over the last 2 decades (1995–2014): Analysis of current data in the United States. Biological Psychiatry 2016; 79(7): 613–619.

Hemant Lata, Suman Chandra and Mahmoud A. ElSohly*

2 *Cannabis*: botany and biomass production

Abstract: The plant cannabis belongs to the family Cannabaceae. It has been cultivated and used as a source of fiber, food, and medicine since ancient times. The plant
produces a unique class of terpenophenolic compounds called cannabinoids. So far,
more than 550 constituents have been isolated from *Cannabis sativa*, out of which 125
are phytocannabinoids. Among phytocannabinoids, Δ^9-tetrahydrocannabinol is reported to be the most psychoactive compound with a wide spectrum of therapeutic
potential. On the other hand, cannabidiol, a nonpsychoactive compound, is reported
to possess very promising pharmacological activities, most notably as an antiepileptic
agent, particularly for the treatment of intractable pediatric epilepsy. In view of the
variation in pharmacological activities of different cannabinoids, there is much interest in growing compound-based crops for research purposes. In this chapter, we ad-
dress cannabis botany and its indoor and outdoor cultivation under good agricultural
practices (harvesting, processing, and storage to be used as a starting material), to develop a botanical drug.

Keywords: Cannabidiol, botanical drugs, cannabinoids, cryopreservation, cultivation,
Δ^9-tetrahydrocannabinol, extraction, hydroponics, phytocannabinoids, micropropagation, propagation

2.1 Introduction

2.1.1 Medicinal and cultivational history

Plants have a long history of being used as drugs/medicine by humans. The knowledge
of plants as medicine is the foundation of many modern medicines and practices. According to the World Health Organization, 80% of the world's population still depend
on plants for their primary healthcare needs, in one form or another. A single compound-based drug originated from plants or a botanical drug (single species or a

Acknowledgment: This work was supported in part by the National Institute on Drug Abuse (NIDA), National Institutes of Health (NIH), Department of Health and Human Services, USA.

*Corresponding author: Mahmoud A. ElSohly,** National Center for Natural Products Research, School
of Pharmacy, University of Mississippi, MS 38677, USA; Department of Pharmaceutics and Drug Delivery,
School of Pharmacy, University of Mississippi, MS 38677, USA, e-mail: melsohly@olemiss.edu
Hemant Lata, Suman Chandra, National Center for Natural Products Research, School of Pharmacy,
University of Mississippi, MS 38677, USA

https://doi.org/10.1515/9783110718362-002

mixture of several plant species) faces unusual challenges during its journey from nature (plant) to pharmaceuticals involving a multifaceted approach combining botanical, phytochemical, biological, and molecular techniques. In the case of cannabis, being a psychoactive substance, a major additional complexity is derived from the regulatory angle with the concerns of its potential of abuse. The plant cannabis has been used as a natural therapeutic herb since ancient times. *Cannabis* cultivation and use is 5,000–6,000 years old, making it difficult to pinpoint the origin of this species [1]. *Cannabis* has a long history of medicinal use in the Middle East and Asia, dating back to the sixth century BC, while it was introduced in Western Europe as a medicine in the early nineteenth century to treat epilepsy, tetanus, rheumatism, migraine, asthma, trigeminal neuralgia, fatigue, and insomnia [2, 3]. *Cannabis* is valued for its hallucinogenic and medicinal properties, having been used to treat a variety of ailments including pain [4], glaucoma [5], nausea [6], depression [7], and neuralgia [8]. The therapeutic value of cannabis and phytocannabinoids has also been employed for HIV/AIDS symptom management [9], in multiple sclerosis treatment [10], and as an antiepileptic agent, particularly for the treatment of intractable pediatric epilepsy [11].

The plant is an annual herb which has been widely dispersed and cultivated by humans in almost all parts of the world. It is one of the oldest plant sources for seed oil, intoxicant resin, medicine, and textile fiber [3, 12]. Archaeological evidences indicate that cultivation of cannabis originated in China for fiber crop and subsequently spread to the Middle East, Europe, and South America during the early sixteenth century [13, 14]. Hemp cultivation in Europe became widespread after 500 C.E. The crop was brought to South America (Chile) in 1545 and North America (Port Royal, Canada) in 1606 [15]. But it is really difficult to pinpoint its original geographical distribution since this species has been spread and modified by humans for thousands of years. Cannabis is currently considered a schedule-I drug, and its cultivation in the United States is highly regulated under the federal law. A detailed description of cannabis history is given in Chapter 1.

2.1.2 Cannabis constituents and cannabinoid biosynthesis

Based on chemistry, botany, and pharmacology of its number of constituents, cannabis is considered as a complex plant. It is the natural source of cannabinoids, mainly accumulated in the glandular trichomes of the plant. These compounds (phytocannabinoids), present in the cannabis plant, are in the acid form and are the precursors of the neutral compounds responsible for the pharmacological activities. Cannabinoid acids undergo decarboxylation process with age or heating to form neutral cannabinoids [16]. For example, Δ^9-tetrahydrocannabinol acid A (Δ^9-THCA) forms Δ^9-tetrahydrocannabinol (Δ^9-THC), and cannabidiolic acid (CBDA) forms cannabidiol (CBD). Chemical structures of THC and CBD are shown in Figure 2.1.

The first cannabinoid to be isolated [17] and identified [18–20] from *C. sativa* was cannabinol (CBN). CBD was subsequently isolated from Mexican marijuana [21], and

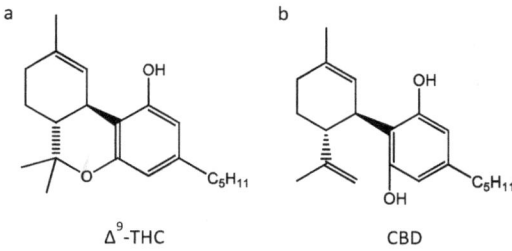

Figure 2.1: Chemical structures of two major phytocannabinoids, responsible for most of the pharmacological effects, present in cannabis: (**a**) Δ^9-tetrahydrocannabinol (Δ^9-THC) and (**b**) cannabidiol (CBD).

the structure was determined by Mechoulam and Shvo [22] in 1963. The most psycho-active compound Δ^9-THC was isolated, and the structure was determined by Gaoni and Mechoulam [23] in 1964. Since then, the number of cannabinoids and other compounds isolated from or identified in cannabis has been continually increasing. Today, more than 550 constituents have been reported in cannabis. Out of which, 129 are phytocannabinoids [24, 25] (Table 2.1). Details about cannabis constituents are given in Chapter 4. Researchers have well documented the pharmacologic and thera-peutic potency of *Cannabis* preparations and Δ^9-THC [26–28]. In more recent days, there has been a significant interest in CBD because of its reported activity as an anti-epileptic agent, particularly its promise for the treatment of intractable pediatric epi-lepsy [29].

Table 2.1: Number of cannabinoids separated by subclasses in *Cannabis sativa* L.

Cannabinoids	1980	1995	2005	Current
CBG type	6	6	7	16
CBC type	4	4	5	9
CBD type	7	7	7	13
Δ^9-THC type	9	9	9	25
Δ^8-THC type	2	2	2	5
CBL type	3	3	3	3
CBE type	5	5	5	5
CBN type	6	7	7	11
CBND type	2	2	2	2
CBT type	6	9	9	9
Misc. type	11	12	14	31
Total	61	66	70	129

CBG, cannabigerol; CBC, cannabichromene; CBD, cannabidiol; THC, tetrahydrocannabinol; CBN, cannabinol; CBL, Cannabicyclol; CBE, Cannabielsoin; CBND, Cannabinodiol and CBT, Cannabitriol.

Besides Δ^9-THC and CBD, other major cannabinoids of *Cannabis* include tetrahydrocannabivarin, cannabichromene, cannabigerol (CBG), and CBN. Δ^8-THC is another closely related isomer of Δ^9-THC, which is much less abundant and less potent than Δ^9-THC [30].

The biosynthesis of cannabinoids has been extensively reviewed by several researchers [31–34] (Figure 2.2). Two independent pathways, the cytosolic mevalonate and the plastidial methylerythritol phosphate (MEP), are responsible for plant terpenoid biosynthesis. The plastidial MEP pathway is reported to lead terpenoid biosynthesis [35]. Formation of olivetolic acid (OLA) is the first step in the cannabinoid biosynthetic pathway. OLA and geranyl diphosphate (GPP) are derived from the

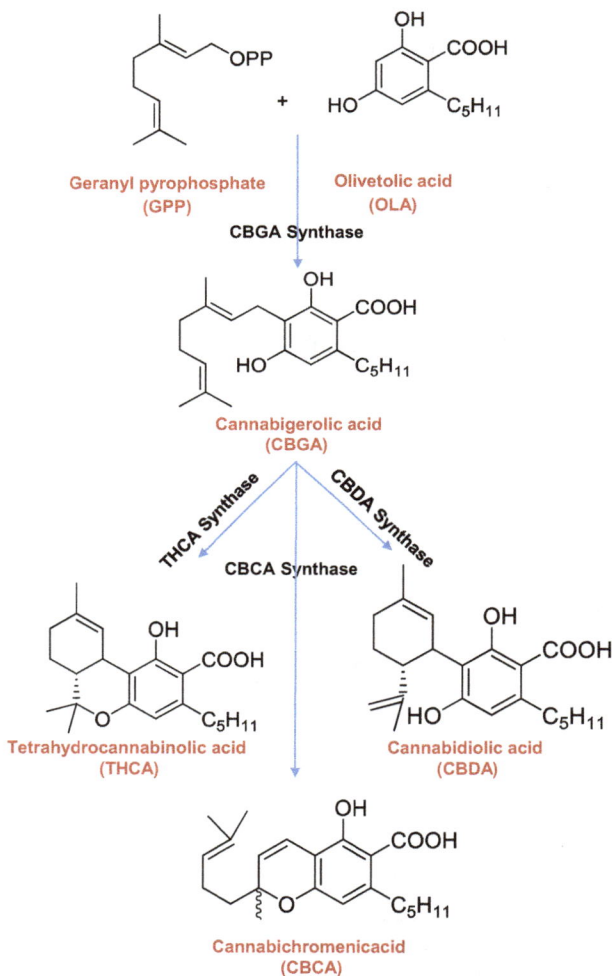

Figure 2.2: Biosynthetic pathway of cannabinoids.

polyketide and the deoxyxylulose phosphate/MEP pathways, respectively. Cannabigerolic acid is obtained by the action of prenylase, GPP:olivetolate geranyltransferase, which is further oxidocyclized by flavin adenine dinucleotide-dependent oxidases, such as cannabichromenic acid (CBCA) synthase, CBDA synthase, and Δ^9-THCA synthase, producing CBCA, CBDA, and Δ^9-THCA, respectively [36]. Detailed biosynthesis of plant cannabinoids is described in Chapter 3.

2.2 Botanical aspects

2.2.1 Nomenclature

The plant cannabis is highly variable and complex in its morphology and genetics; therefore, the number of species in the *Cannabis* genus is a matter of debate for a long time. Modern hybrid varieties and/or cultivars with manipulation in genetics and chemical constituents make the nomenclature of cannabis more difficult. The pioneer of modern taxonomy, Swede Carl Linnaeus [37], treated cannabis as a single species, whereas Lamarck [38] described "Indian cannabis strain" as taxonomically different than "European hemp" and gave it a specific name *Cannabis indica*. The botanical nomenclature of cannabis includes **kingdom**: Plantae (plants); **subkingdom**: Tracheobionta (vascular plants); **superdivision**: Spermatophyta (seed plants); **division**: Magnoliophyta (flowering plants); **class**: Magnoliopsida (dicotyledons); **subclass**: Hamamelididae; **order**: Urticales; **family**: Cannabaceae; **genus**: *Cannabis*; and **species**: *Cannabis sativa* L.

In 1976 [39], Small and Cronquist reported cannabis as a single species (monospecific concept) and have divided it into *Cannabis sativa* L. subsp. *indica* (L.) and *C. sativa* L. subsp. *sativa*, whereas Hillig [40], on the basis of allozyme data, showed that cannabis has been derived from two major gene pools, and on the basis of this data, he recognized *C. sativa* and *C. indica* as separate species. *Cannabis ruderalis*, a wild cannabis, native to Eastern Europe and Russia with low THC, was also considered a separate species from sativa and indica.

The taxonomic disagreement still revolves around how to assign scientific names to different *Cannabis* strains with different morphological and chemical profiles, specifically the modern hybrid varieties. Based on the morphological and phenotypic variations, a number of reports proposed *Cannabis* as a polytypic genus, whereas others suggest as monotypic but highly polymorphic species, *C. sativa* L. [41–47]. At the present day, based on morphological, anatomical, phytochemical, and genetic studies, *Cannabis* is generally treated as a single but highly polymorphic species, *C. sativa* L., having different varieties such as *C. sativa* var. *sativa*, *C. sativa* var. *indica*, and *C. sativa* var. *ruderalis*.

On the other hand, in view of the highly hybridized and manipulated modern cannabis varieties on the market, it is very difficult to categorize cannabis based on morphophenological characters. Therefore, cannabis varieties are also characterized based on their cannabinoid and terpenoid profiles (and contents) for more clarity and scientific accuracy [48].

The plant has many local common names, and some of which are given in different languages. Cannabis in **Arabic** is known as bhang, hashish, qinnib, hasheesh, kenneb, and qinnibi tîl; in **Chinese** as xian ma and ye ma; **Danish** as hemp; **Dutch** as hennep; in **English** as hemp, marihuana, and marijuana; **Finnish** as hamppu; **French** as Chanvre, chanvre d'Inde, chanvre indien, and chanvrier; **German** as hanf, haschisch, and indischer hanf; **Hindi** as bhang, charas, and ganja; **Japanese** as Mashinin; **Nepalese** as charas, gajiimaa, and gaanjaa; **Portuguese** as cânhamo and maconha; **Russian** as Kannabis sativa; **Spanish** as cáñamo, grifa, hachís, mariguana, and marijuana; and in **Swedish** as porkanchaa.

2.2.2 Botanical identification

Cannabis, as the crude drug, consists of the flowering top of the female plants of *C. sativa* L. The plant is an annual herb with generally dioecious (i.e., female and male flowers found on separate plants) and rarely monoecious flowers (i.e., male and female flowers found on the same plant, commonly referred to as hermaphrodite).

They have palmately compound leaves with serrate margins, consisting typically of 5 or 7, occasionally 9 or 11, and rarely more than 11 leaflets. The flowering top consists of ascending branches, all longitudinally furrowed and bearing numerous bracts, and in the axil of which small cymose inflorescence is nested. The bract may be simple and sessile, or palmately compound with three leaflets and shortly petiolate, each having two small stipules; the laminae are lanceolate and have entire margins and are usually 1.5–2.0 cm long and 0.2–0.4 cm wide.

The lower bracts are somewhat larger and resemble the foliage leaves; they are palmately compound leaves with three to five leaflets that are narrowly lanceolate, with serrate margins and acute apices, and the central and largest leaflet is about 3 cm long. In the axil of each bract are two boat-shaped bracteoles with acute apices, each enclosing a single pistillate flower. The flower consists of an ovary enveloped by a membranous perianth, termed as perigone, and the ovary is about 2 mm long and is surmounted by two long brownish-red stigmas; within the ovary, there is a single ovule.

The fruit is 5–6 mm long and 4 mm wide, ovoid with several longitudinal veins due to the presence of the enlarged, persistent bracteole enveloping the fruit. The seed fills the loculus and contains a large embryo and very scanty endosperm. All parts of the plant, but particularly the bracts, stipules, and upper leaves, bear numerous trichomes (hairs). Microscopically, a transverse section in the leaves and bracts has the structure of a dorsiventral leaf.

The palisade consists of a single layer, rarely two, of cylindrical cells and the spongy tissue of two to four layers of rounded parenchyma; cluster crystals of calcium oxalate are present in all parts of the mesophyll. The upper epidermis has cells with straight anticlinal walls and bears unicellular, sharply pointed, curved conical trichomes, about 150–220 µm long, with enlarged bases having cystoliths of calcium carbonate; the lower epidermis bears conical trichomes which are longer, about 340–500 µm long, and more slender, but without cystoliths. Both upper and lower epidermis bear numerous glandular trichomes, especially over the midrib.

The glandular hairs have a long multicellular pluriseriate or short unicellular stalk and multicellular head with about eight radiating club-shaped cells. Stomata of the anomocytic type (i.e., the stoma is surrounded by a varying number of cells that do not differ from those of the epidermis generally) are present on the lower surface of the leaf but are absent from the upper surface. The bracteoles have an undifferentiated mesophyll of about four layers of cells, the lower hypodermal layer having a cluster crystal of calcium oxalate in almost every cell. The abaxial surface bears numerous glandular trichomes and also unicellular conical trichomes, which are numerous on the bulging part, but scattered thinly elsewhere. The stigmas have an epidermis nearly every cell which extends as a unicellular papilla of about 90–180 µm long with a rounded apex.

The slender stem axis has well-developed bundles of pericyclic fibers behind the phloem bundles. There are large laticiferous tubes in the phloem and calcium oxalate in cluster crystals, about 25–30 µm in diameter, in both the pith and the cortex. The epidermis bears very few trichomes similar to those of the leaves. The laticiferous tubes are unbranched.

Cannabis seeds germinate in 3–7 days. The first pair of leaves usually has a single leaflet, the number gradually increasing up to 11 leaflets per leaf and rarely more (usually 5–9), depending on the variety and growing conditions. *Cannabis* can grow to a height of 5–7 m in a 4- to 6-month growing season. Under favorable conditions, it can grow (vegetative stage, Figure 2.3A) up to 10 cm in height in a given day during the long summer days. However, the growth of the plant slows down during the later (flowering, Figure 2.3B) part of the season.

Stems of this plant are green, erect, sometimes hollow, and longitudinally grooved. Most strains of *Cannabis* are short day plants and flower when exposed to a photoperiod of less than 12 h. Cannabis is a dioecious plant, with staminate (male) and pistillate (female) flowers occurring on separate plants. However, it is not unusual for individual plants to bear both male and female flowers.

Monoecious *Cannabis* plants are often referred to as "hermaphrodites," and bear male and female flowers at different locations on the same plant. Cannabis, in general, is described as a dioecious plant. It is difficult to identify male plants from females at the vegetative stage. However, after the onset of flowering (based on flowering pattern), male and female plants can easily be discriminated. Male flowers are normally borne on loose panicles and female flowers are borne on racemes.

Figure 2.3: (A) Vegetative and (B) flowering stages of cannabis plant.

2.2.3 Macroscopic identification

Cannabis is an annual herb. Morphological characteristics and variation in color are influenced by the seed strain and by environmental factors such as light, water, nutrients, and space. As a dioecious herb, the flowers on individual plants of cannabis are unisexual; however, the plant some time exhibits monoecious (flowers of the opposite sex on the same plant). Morphology of *C. sativa* plant is shown in Figure 2.5.

Male plants (Figures 2.4A and 2.5A) are usually taller, less robust, and flower earlier than female plants. Stems are green, erect, sometimes hollow, and longitudinally grooved. They can vary from 0.2 to 6 m, although most of the plants reach heights of 3–4 m during regular outdoor cultivation. The extent of branching and plant height depends on environmental and hereditary factors as well as the method of cultivation. The side branches vary from opposite to alternate at any part of the main stem. The leaf arrangement changes from decussate (oppositely arranged) to alternate on the extremities of the plant.

Leaf stalks (petioles) are 2–7 cm long with a narrow groove along the upper side. The leaf is palmate and consists of 1–14 leaflet blades of 3–15 × 0.2–1.7 cm. The margins are coarsely serrated, the teeth pointing toward the tips, and the veins run out obliquely from the midrib to the tips of the teeth. The lower (abaxial) surfaces are pale green with scattered, white to yellowish brown, resinous glands. Each staminate (male) flower consists of five whitish-green minutely hairy sepals about 2.5–4 mm long and five pendulous stamens, with slender filaments and stamen.

Figure 2.4: (A) Male and (B) female plants (flowers) of *Cannabis sativa.*

The pistillate (female) flowers (Figures 2.4B and 2.5B) are more or less sessile and are borne in pairs. Each flower has a small green bract enclosing the ovary with two long, slender stigmas projecting well above the bract. The fruit, an achene, contains a single seed with a hard shell tightly covered by the thin wall of the ovary, ellipsoid, slightly compressed, smooth, about 2–5 mm long, generally brownish, and mottled. The fruit is commonly regarded as a seed.

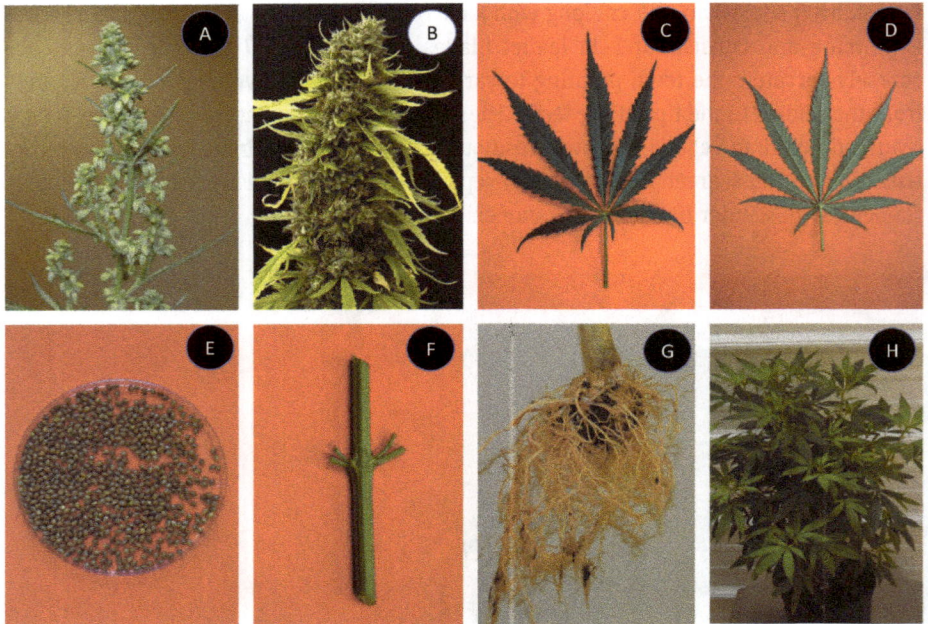

Figure 2.5: Morphology of *C. sativa*: (A) Male and (B) female flower tops; (C) adaxial and (D) abaxial leaf surfaces; (E) seeds; (F) stem; (G) roots; and (H) a fully developed cannabis plant.

2.2.4 Microscopic identification

2.2.4.1 Based on trichomes

Cannabis can be identified by microscopic trichomes, that is, hair-like projections from a plant's epidermal cell, on the surface of the plant. These trichomes can be observed with a microscope at 40× magnification (Figure 2.6).

Based on their formation/structures, trichomes are categorized into two distinct classes:

(a) Glandular trichomes
(b) Nonglandular trichomes

The glandular trichomes produce and store the cannabinoid-containing resin. They are mainly associated with the flower structures (pistillate plants being particularly rich in these structures), but they can also be found on the underside of the leaves and occasionally on the stems of young plants. Glandular trichomes are divided into three different categories: (1) glandular stalked, (2) glandular sessile, and (3) bulbous glandular trichomes.

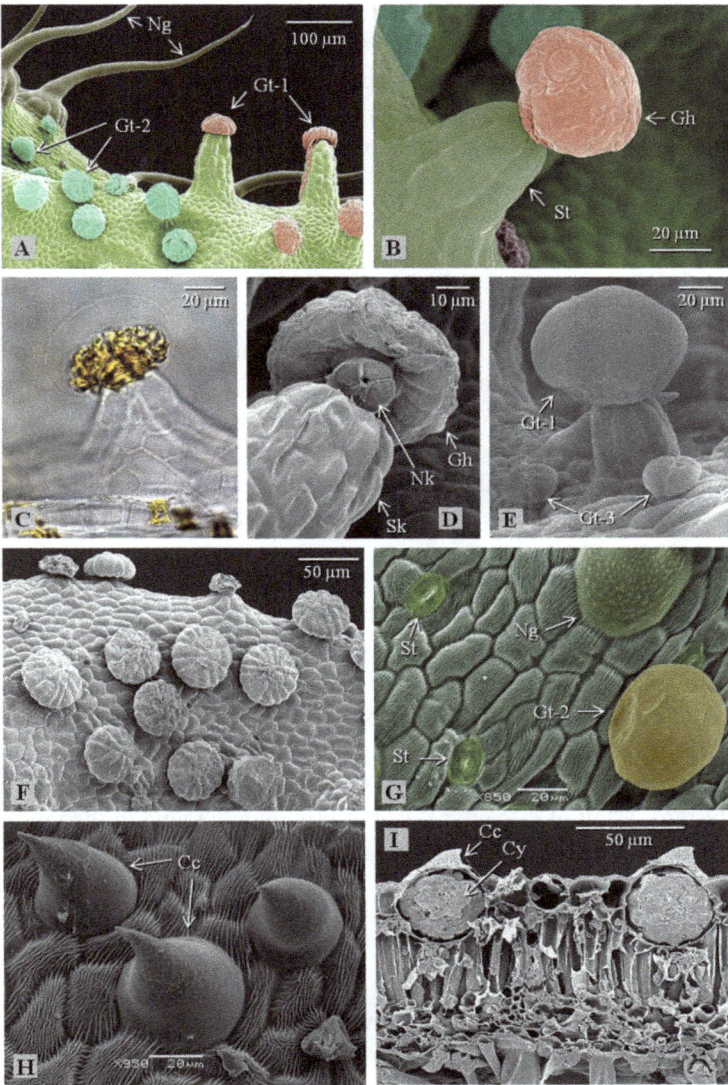

Figure 2.6: Glandular and nonglandular trichomes in *C. sativa* (C: LM; all other SEM; A, B, G: colorized SEM images). (A) Portion of bract displaying glandular and nonglandular trichomes; (B–D) capitate stalked glandular trichomes (note an "eyespot" on the glandular head in image (B); the glandular disk and cuticular membrane in (C); and a slightly broken "neck" of glandular head showing four-cell arrangement, in image (D); (E) a capitate stalked glandular trichome and two of the bulbous glandular trichomes; (F) a group of capitate sessile glandular trichomes on a young leaf; a sessile glandular trichome on abaxial leaf surface (note the presence of stomata); (H) morphology of conical cystolith trichomes on adaxial leaf surface; and (I) cystolith trichomes in sectional view showing large cystolith crystals.
Gh, glandular head; Gt-1, capitate stalked glandular trichome; Gt-2, capitate sessile glandular trichome; Gt-3, bulbous glandular trichome; Ng, nonglandular trichome; Nk, "neck"; Sk, stalk; St, stomata.

Nonglandular trichomes are unicellular, rigid, curved hairs with a slender pointed apex and are of two major types: (1) cystolithic and (2) noncystolithic trichomes.

Cystolithic trichomes found on the upper surface of the leaves have a characteristic bear claw shape and may have calcium carbonate crystals (cystoliths) visible at their bases. Noncystolithic trichomes occur mainly on the lower side of the leaves, bracts, and bracteoles, and lack the enlarged base.

2.2.4.2 Based on anatomical characteristics of cannabis leaf

Anatomical characteristics of cannabis leaf are shown in Figure 2.7. In transection, the leaf of *C. sativa* shows thin lamina and major veins, which are depressed above and prominently raised beneath. Each of the upper and lower epidermis is unilayered. In surface view, the epidermal cells show slightly undulate anticlinal walls. Upper epidermis shows the characteristic cystolith trichomes with an enlarged base containing large cystolith crystal. Numerous nonglandular and glandular trichomes are present on the lower epidermis.

Stomata are numerous on the lower epidermis and are not observed in the upper epidermis. The mesophyll consists of palisade and spongy tissue. Palisade is unilayered, consists of thin columnar cells, and occupies more than half thickness of the lamina. Spongy cells are loosely arranged with large air spaces, leading to stomatal cavities. Transection of midrib (Figure 2.7) shows a single collateral vascular bundle. Small groups of collenchyma cells are present beneath the upper epidermis and inside the lower epidermis. A few laticifer ducts with yellow-brown secretions are found in the phloem [49]. Cluster crystals of calcium oxalate are commonly found in the mesophyll and phloem parenchyma of the veins [50].

2.2.5 Cannabis: drug and fiber types

Based on morphology and with plant growth and development, cannabis is divided into two distinct categories: (a) drug-type and (b) fiber-type plants. Drug-type varieties have more robust growth and grows like a Christmas tree with big branches at the lower part of the stem, whereas fiber-type varieties grow skinny and tall with very few branches.

Cannabis is considered a chemically complex species based on its numerous natural constituents [51]. The concentration of Δ^9-THC in dried inflorescence (leaves and buds) is used to determine its psychoactivity. Quantitative and qualitative analyses of

---

Figure 2.6 (continued)
Reprinted from "Chandra S, Lata H, Elsohly MA (Eds.) *Cannabis Sativa* L.: Botany and Biotechnology, Springer Cham, 2017" with permission from Springer Nature.

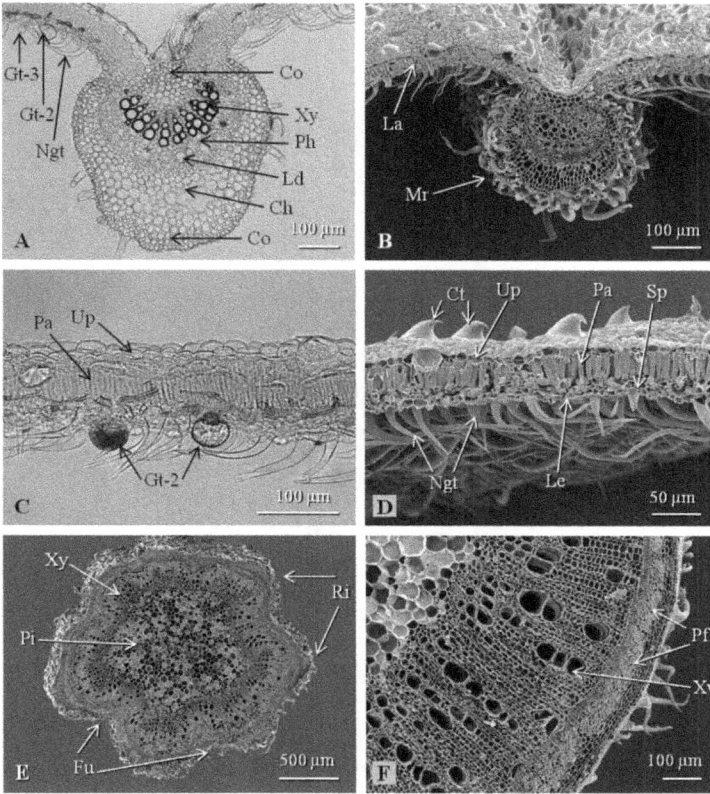

Figure 2.7: Anatomy of *C. sativa* (A and C: LM; all other SEM). (A, B) Transection (TS) of leaf through midrib; (C, D) TS of leaf through lamina; (E) TS of stem, with a portion enlarged (F).
Ch, chlorenchyma; Co, collenchyma; Ct, cystolith trichome; Fu, furrows; Gt-2, capitate sessile glandular trichome; Gt-3, bulbous glandular trichome; La, lamina; Ld, laticifer duct; Le, lower epidermis; Mr, midrib; Ngt, nonglandular trichome; Pa, palisade; Pf, pericyclic fibers; Ph, phloem; Pi, pith; Ri, ridges; Sp, spongy tissue; Up, upper epidermis; Xy, xylem.
Reprinted from "Chandra S, Lata H, Elsohly MA (Eds.) *Cannabis Sativa* L.: Botany and Biotechnology, Springer Cham, 2017" with permission from Springer Nature.

cannabis can be employed to characterize their phenotype and phytocannabinoid profile [52]. Cannabis can be divided into three distinct phenotypes:

Phenotype I (drug type), with Δ^9-THC >0.5% and CBD <0.5% (Δ^9–THC/CBD $\gg$ 1) [53].

Phenotype II (intermediate type), with CBD as the major cannabinoid but with Δ^9-THC also present at various concentrations (Δ^9-THC/CBD ~ 1) [54].

Phenotype III (fiber type or hemp), with especially low Δ^9-THC content (Δ^9-THC/CBD $\ll$ 1) [53].

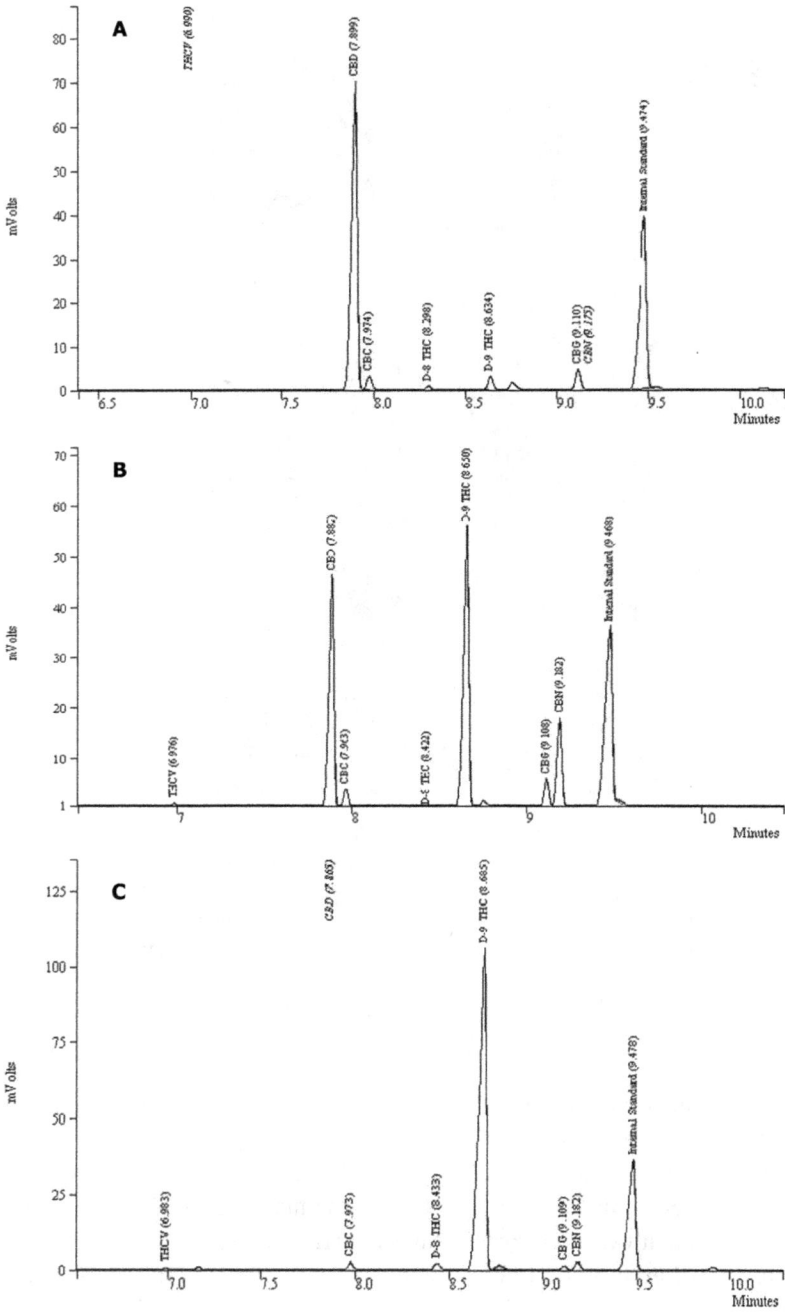

Figure 2.8: (A) Fiber-type cannabis (CBD > THC), (B) intermediate-type cannabis (THC ~ CBD), and (C) drug-type cannabis (THC > CBD).

Representative GC-FID (gas chromatography-flame ionization detection) chromatograms of drug-type, intermediate-type, and fiber-type cannabis biomass samples are shown in Figure 2.8.

A rare additional chemotype, characterized by a higher concentration of CBG and very low concentration of CBD and THC, was reported by Fournier et al. [55].

Hemp is usually described as a nonpsychoactive cannabis plant with predominant CBD or CBG contentment [56]. From the regulatory point of view, cannabis containing <0.3% THC is considered as hemp, whereas cannabis plant containing >0.3% THC is considered as drug-type (controlled) substance.

Although environmental factors play a role in the number of cannabinoids present in different parts of the plant at different growth stages [57], the distribution of CBD:Δ^9-THC ratios in most populations is under genetic control [58]. The phytocannabinoid content of cannabis is determined by the interaction of several genes, cultivation technique, and environmental factors [59–63]. Numerous biotic and abiotic factors affect phytocannabinoid production, including the sex and maturity of the plant [39], light cycle [64], temperature [59], fertilization [57], and light intensity [59]. Variations in phytocannabinoid content in different tissues for a specific plant have also been reported [62].

In recent years, several studies have shown that molecular techniques (using genomic DNA) such as "random amplified polymorphic DNA" and "restriction fragment polymorphisms" analysis can efficiently be used to differentiate cannabis varieties [65] or can be used to identify genetic relationship among different varieties of *Cannabis* plants.

2.2.6 Cannabis: male and female plants

Cannabis is a dioecious plant, which means, if grown from seeds, a single seed turns into either a male or a female plant (Figures 2.4 and 2.5A and B). Occasionally, cannabis plant shows monoecious nature, which means male and female, and both types of flowers appear on different branches of a single plant.

Cannabis flowers under the shorter photoperiod (below 12 h photoperiod, shorter days) and continues growing vegetatively under the longer (photoperiods) days. At the vegetative stage, it is difficult (almost impossible) to differentiate between male and female plants due to their morphological similarities. However, separation of male from female is possible at the early flowering stage based on their different morphological appearance. However, in the last few years, there are reports describing techniques to discriminate between male and female cannabis plants at the early growth (vegetative) stage using molecular markers [66].

Cannabis is a wind-pollinated plant with allogamous (cross-pollination or fertilization) nature. If grown from seeds (indoor or outdoor), it is difficult to maintain chemical profile of a cannabis biomass product, especially with other varieties. In the modern horticulture, "sinsemilla" (a Spanish term for seedless plants) cannabis plants are preferred over "seedlings" for the production of research/pharmaceutical-grade biomass for several reasons.

First, in the presence of male plants, female cannabis plants produce seeds, and therefore, less inflorescence (leaves and buds), whereas maximum biomass production is of utmost goal of any cultivation setup.

Second, if the crop is grown through seeds in open fields and male plants are removed as they appear, almost 50% field area becomes empty unless the space is filled by female plants grown through cuttings or seeds.

Third, and most important, if seedlings of several cannabis varieties are being grown together, in the presence of male plants, cross-pollination occurs among varieties that affect the genetic makeup of harvested seeds and biomass, resulting in subsequent product (seeds and biomass) inconsistent in its cannabinoid profile and content from the original batch.

Batch-to-batch consistency in chemical profile is an important factor for any pharmaceutical-grade biomass product. In cannabis production, this can be achieved by screening and selection of mother clones with desired chemical profile, their conservation, and cultivation to produce a biomass product consistent in chemical profile.

2.2.7 Screening and selection of mother clones

A schematic diagram of screening and selection of female mother clones is shown in Figure 2.9. As described earlier, cannabis is a highly polymorphic plant and is very difficult to maintain consistency in the final biomass product if grown from seeds. Plant-to-plant variations in terms of cannabinoid content are reported in seed-raised cannabis biomass, even though seeds are obtained from a single plant. To avoid this phenomenon, seeds are grown individually in small jiffy pots under the climate-controlled indoor growing environment. Once germinated and well rooted, plants are transferred to bigger pots for the vegetative growth (18 h photoperiod). After achieving the desired vegetative growth, few "well-developed healthy plants (mother plants)" are selected for the screening (these plants are always maintained under the vegetative light environment throughout the screening). Each plant is given a unique identification number. Cuttings are made from each mother plant, and brought to a desirable vegetative (18 h photoperiod) growth and then subjected for flowering (12 h photoperiod). On flowering, all the male cuttings and their related "mother plants" are removed from the growing room, and the female cuttings at flowering are grown till the formation of mature buds. Biomass samples from mature buds of each plant are analyzed for cannabinoid content using GC-FID. Based on cannabinoid analysis, high-yielding candidates are identified and their related mother plants are selected as elite "mother plants" for future propagation.

Selection of seeds of desirable *Cannabis sativa* variety
↓

Seeds were germinated
↓

Seedlings were grown under the vegetative light environment
↓

Among the seedlings, few 'well-developed healthy plants (mother plants)' were selected for the screening (these plants are always maintained under the vegetative light environment throughout the screening). Each plant was given a unique identification number
↓

Cuttings were made from each mother plant
↓

These cuttings were raised up to a desirable vegetative (18h photoperiod) growth and then subjected for flowering (12h photoperiod)
↓

On flowering all the male cuttings and their related 'mother plants' were removed from the growing room
↓

Female cuttings were grown till the formation of mature buds. Biomass samples from mature buds of each plants were analyzed for cannabinoids content using GC-FID
↓

Based on cannabinoids analysis, high yielding candidates were identified and their related mother plants were selected as elite 'mother plants' for the future propagation.

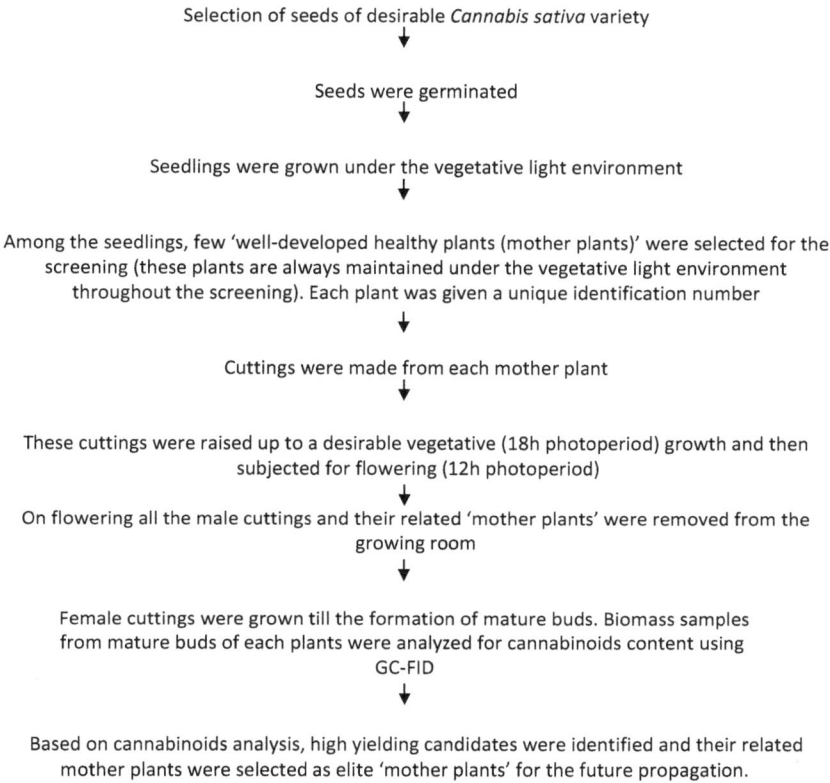

Figure 2.9: Screening and selection of elite mother plant of *Cannabis sativa* for the propagation of true to type plants.

2.3 Biomass production

C. sativa L. is an annual herb, which means it grows vegetatively during long summer days and flowers and make buds during short days as the winter season approaches. The plant eventually dies if not harvested at the peak flowering stage. Cannabis can be grown "indoor" and "outdoor" efficiently. However, each cultivation option has its advantages and disadvantages. Under the outdoor conditions, the life cycle of the plant completes in 5–7 months depending on the time of plantation and the variety, whereas growing indoor, flowering can be triggered by regulating the photoperiod. Outdoor cultivation is affected by factors such as wind and rain that can destroy the cannabis plants. Other environmental variables such as temperature, light, water availability, and plant spacing also affect the growth and development of cannabis plants, causing variations in quantity and quality of biomass.

2.3.1 Indoor cultivation

Indoor cultivation of *C. sativa* under controlled environmental conditions provides many benefits compared to outdoor cultivation, including total control on "plants life cycle at different stages of growth" and "the quality," and "quantity" of the biomass product. The following parameters are to be considered for indoor production:

2.3.1.1 Environmental parameters

The goal of indoor growth is to produce a "biomass product" consistent in chemical profile and content. This is achieved through providing optimum environmental conditions to the plants through their lifecycle. A strict control on the following environmental parameters of the grow room is critical for effective cultivation of cannabis plants and to avoid pest and diseases.

Photosynthetic photon flux density (PPFD) and photoperiod: Optimum light quality (wavelengths, photosynthetically active radiation, ranging from 400 to 700 nm), quantity (PPFD, $\mu mol/m^2$ s), and photoperiod (light hours) are the most important factors to maximize biomass production of an indoor grown crop. Light quality and quantity have a profound influence on photosynthesis which ultimately affects the plant growth and development [67]. Optimum light quality and quantity provided to plants make them bushier with shorter distance between the nodes. Cannabis, in particular, is reported to be benefited from high PPFD for photosynthesis and growth [59]. Different light sources can be used for indoor propagation such as fluorescent light bulbs (mainly for young cuttings), metal halide bulbs, high-pressure sodium lamps, induction bulbs, and light-emitting diodes (LEDs). LEDs are meant to produce less heat as compared to metal halide and sodium lamps. To avoid overheating, a safe distance is maintained between bulbs and plants. A photoperiod of 18 h or more is desirable for vegetative growth, whereas 12-h photoperiod is recommended for the initiation of flowering.

Temperature: Temperature dependence of photosynthesis is reviewed by several authors in different plant species [68–71]. In cannabis, plant development and growth of different varieties vary depending upon their original growth habitat and the genetic makeup. In general, 25 °C growth temperature is found to be optimum for most varieties of cannabis [59, 72]. However, it is important to maintain an ideal combination of light intensity and grow room temperature for optimum growing. High light intensity with high temperature is reported to have an adverse effect on the photosynthesis of cannabis plants [59].

Irrigation and relative humidity: Cannabis requires high humidity at the juvenile (cutting or seedling) stage, whereas a comparatively drier environment at the flowering stage is needed. Humidity plays a critical role at every stage of cannabis plant. In a close grow room environment, accumulation of humidity or moisture is quite common due to irrigation and water evaporated by plants. Proper ventilation, air circulation, and sometimes dehumidification are required to maintain optimum conditions. The amount of water and the frequency of watering of cannabis plants vary with the growth stage, size of the plants and containers, growth temperature, humidity, and many other factors. Vegetative cuttings require a regular water spray on the leaves to maintain a high humidity in its microclimate until the plants are well rooted. Once established, the top layer of soil must be allowed to dry out before the plants are watered again. Humidity around 75% is recommended during the juvenile stage and about 50–55% during the active vegetative and flowering stages.

Carbon dioxide concentration: Constant air circulation, proper ventilation of air, and drier grow room environment can prevent mold formation and several plant's related diseases. Air flow around the plants and a steady fresh air flow from outdoor are recommended in indoor grow room for healthy cannabis crop.

Several studies suggest that elevated CO_2 causes increased photosynthesis in plants [73–77]. A close correlation between photosynthesis and plant biomass yield is reported by Zelitch [78]. Doubling in CO_2 concentration has been reported to increase the yield by 30% or more in many crops [79]. In *C. sativa*, doubling of CO_2 concentration (~750 ppm) was reported to stimulate the rate of photosynthesis in different varieties by 38–48% as compared to ambient CO_2 concentration [80]. Therefore, supplementing CO_2 to the existing amount in the grow room during the light cycle is recommended for vigorous cannabis growth.

2.3.1.2 Propagation through seeds and vegetative cuttings

For cultivation of *Cannabis*, seeds have been the main source of propagation. Well-aerated and moist soil is preferred for sowing seeds in small jiffy pots. If the seed bed is outdoor and the nights are cold, heat mat can be used below the pots to increase the temperature. The seedling should begin to sprout by the fourth day, and most of the viable seeds germinate by 2 weeks.

Variation in the rate of seed germination depends on the variety, seed age, storage condition, and soil and water temperatures. Cannabis cultivated outdoors need full sunlight to grow profusely, mature properly, and produce high resin content. Germinated seedlings can be kept under cool fluorescent light with 18-h photoperiod till the seedlings are big enough to transplant in bigger pots. These pots can be kept under full-spectrum grow light (18-h vegetative photoperiod). After enough vegetative growth, plants may be exposed to 12-h photoperiod for flowering. Onset of flowering normally occurs in 2 weeks. At this stage, male plants can be identified. Since male flowers appear before female flowers, male plants can be immediately separated from the female plants if sinsemilla buds are to be produced. Cuttings can be taken from the vegetative branches of selected high-yielding female plants (based on GC-FID analysis) and can be kept under vegetative stage for future propagation.

Vegetative propagation is also referred to as cloning. The technique of growing plants from cuttings from a selected mother plant is a great way to generate crop of consistent quality and reproducibility. Once a particular clone is screened and selected, a fresh nodal segment about 6–10 cm in length containing at least two nodes from the mother plant can be used for vegetative/conventional propagation either in solid (soil) or in liquid medium (hydroponics). Micropropagation, another in vitro vegetative propagation technique, can also be used for the mass propagation of *C. sativa* [81–92].

2.3.1.3 Propagation through hydroponics or aeroponics

Hydroponics and/or aeroponic techniques can be used to grow cannabis plants outdoors (Figure 2.10) as well as in indoor growing (Figure 2.11) conditions. There is a minor difference between the two techniques. In hydroponics, plant roots are constantly submerged in the balanced nutrient solution, whereas in aeroponic technique, roots are misted periodically with nutrient solution. Rock wool and hydrotones (clay balls) are used as a support medium for its growth.

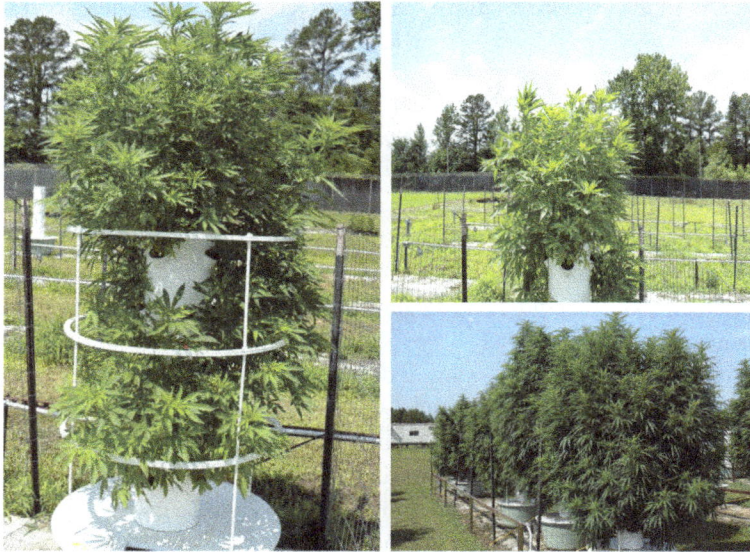

Figure 2.10: Hydroponic cultivation of *Cannabis sativa* under outdoor conditions.

Figure 2.11: Indoor hydroponic cultivation of *Cannabis sativa* under controlled environmental conditions.

A small branch consisting of a growing tip with two or three leaves is cut and immediately dipped in distilled water. Prior to dipping the cutting in a rooting compound, a fresh cut is made just above the first cut. The cuttings are inserted 1 inch deep into a rockwool cube or a hydrotone clay ball supporting medium. Plants are supplied with a vegetative fertilizer formula and exposed to a diffused light:dark cycle (18:6) for vegetative growth. Rooting initiates in 2–3 weeks, followed by transplantation to a bigger hydroponic system.

2.3.1.4 Micropropagation

In vitro culture is one of the key tools of plant biotechnology, based on the concept of totipotency: the ability of plant, cells, and tissues capable of developing into a complete organism or differentiating into any cell types of an organism [93]. It has been used as a tool for the propagation of genetically superior clones, somaclones, ex situ conservation of valuable germplasm, and synthesis of many secondary compounds [94]. The technique reduces the labor and space requirements for producing new varieties of desired characteristics. Micropropagation, defined as the "true-to-type" propagation of a selected genotype is a proven system of regeneration, increasingly being used for *Cannabis* propagation. It has been used as an important aid to conventional methods of propagation. Since *Cannabis* is highly heterozygous and outcrossing in nature, clonal propagation is essential in maintaining genetic fidelity [95]. Moreover, the recent legalization of medicinal cannabis in many countries has opened the doors for cannabis industry, which has immensely applied in vitro propagation approach for multiplication of elite superior varieties. The goal of cannabis micropropagation is to obtain many genetically identical, physiologically uniform, and developmentally normal plantlets in a reduced time and at a lowered cost. Although several micropropagation protocols have been published over the last decade in vitro propagation of cannabis is still challenging. One of the limiting factors in the tissue culture of cannabis variants is associated with the use of different strains, making developed protocols more difficult to establish when another cultivar is being micropropagated [96]. Moreover, it is known that success of in vitro cultivation strongly depends on the genotype, explant type, and phytohormones used in the culture medium [97]. Furthermore, donor conditions of the plant, the age, and physiological state of the parent plant contribute to the success of organogenesis in cell culture facing challenges, where organogenesis responses were difficult to predict even when the same medium was being used [98].

In our laboratory at the University of Mississippi we have developed several protocols to micropropagate cannabis germplasm using apical segment (direct organogenesis) [88, 99] (Figure 2.12), leaf disk (indirect organogenesis) [91], nodal segment in slow growth medium [100], and nodal segment in sodium alginate gels (synthetic seeds) [101], for the conservation and further cultivation [102].

Figure 2.12: Micropropagation of *Cannabis sativa* L. A: *In vitro* rooted plants; B, Acclimatization under climatic controlled environment and C: Fully developed cannabis plants.

2.3.1.4.1 Organogenesis through callus formation (indirect organogenesis)

The hypothesis of organogenesis differentiation by Torrey [103] is an outcome of the process of dedifferentiation followed by redifferentiation of cells. Dedifferentiation favors unorganized cell growth known as callus that is comprised of irregularly differentiated vacuolated cells interspersed with smaller meristematic cells. In our laboratory, we have developed a high-frequency plant regeneration system from leaf tissue-derived callus of a high-yielding variety of *C. sativa* [91, 92]. The concentration and combination of plant growth regulators are the key factors influencing indirect shoot organogenesis in *C. sativa* cultures. The calli were introduced from leaf explants on Murashige and Skoog (MS) medium supplemented with different concentrations (0.5, 1.0, 1.5, and 2.0 µM) of IAA (indole-3-acetic acid), IBA (indole-3-butyric acid), NAA (naphthalene acetic acid), and 2,4-dichlorophenoxyacetic acid in combination with 1.0 µM TDZ (thidiazuron) for the production of callus. The optimum callus growth and maintenance were in 0.5 μM NAA plus 1.0 μM TDZ. The 2-month-old calli were subcultured in MS media containing different concentrations of cytokinins (BAP, kinetin (KN), and TDZ). The rate of shoot induction and proliferation was highest in 0.5 μM TDZ. Regenerated shoots rooted best on half-strength MS medium supplemented with 2.5 μM IBA of various auxins (IAA, IBA, and NAA) tested. Although organogenesis through indirect process offers opportunity to uncover natural variability in plants, regeneration through indirect shoot organogenesis has been shown to produce more somaclonal variants leading to genetic instability of micropropagated plants [104, 105].

2.3.1.4.2 Emergence of adventitious organs directly from explant (direct organogenesis)

Although different routes are available for plant tissue culture regeneration, direct organogenesis is a common method of micropropagation that involves tissue regeneration of adventitious organs or axillary buds directly or indirectly from the explants. Direct organogenesis holds advantages including less culture stages (no callus stage), less or no chances of somaclonal variations, and allowing recovery of genetically stable and true-to-type progeny. We have successfully established direct organogenesis protocol for *C. sativa* using nodal segments.

The explants (nodal segments containing axillary buds) were collected from screened and selected high-yielding *C. sativa* clone grown in an indoor cultivation facility housed at the Coy-Waller Laboratory, University of Mississippi. The explants were cut into small pieces as *1 cm in length and were then treated with 0.5% NaOCl (15% v/v bleach) and 0.1% Tween 20 for 20 min and later washed in sterile distilled water three times for 5 min prior to inoculation on the culture medium. Out of different concentrations of various growth regulators (benzyladenine, BA;KN; and TDZ) tested, the quality and quantity of shoot regenerants in cultures were better with 0.5 μM TDZ. Elongated shoots when transferred to half-strength MS medium supplemented with 500 mg/L activated charcoal and 2.5 μM IBA (as compared to different concentrations of IAA and NAA) resulted in highest rooting. This two-step regeneration protocol utilizes more than one type of growth regulators, that is, TDZ for shoot formation and multiplication, and IBA for rooting [88]. Role of action of TDZ for direct organogenesis has been evaluated by many researchers. Laloue and Pethe [106] proposed that TDZ influences the metabolism of endogenous auxins, thus altering the auxin to cytokinin ratio within the tissue, and eventually stimulating regeneration, and has been supported by several other researchers [107, 108]. TDZ appears to stimulate cells in the apical meristem to divide and multiply for bud differentiation to occur [109]. Thinh [110] suggested that TDZ either increases the levels of nucleoside or the accumulation and synthesis of purine cytokinins as well as promoting the conversion of adenine to adenosine.

We have further improved and refined the existing protocol from two steps to one step for the mass propagation of *C. sativa*. This one-step regeneration protocol is based on the adventitious shoot induction as well as an effective rooting using novel aromatic cytokinin, meta-topolin (mT) [111]. The best response in terms of explants producing maximum number of shoots with maximum shoot length and percent explants producing shoots was recorded on MS medium supplemented with 2 µM mT. Shoots multiplied in the same medium for two subcultures were able to induce healthy roots within 4–6 weeks. mT (an aromatic cytokinin), identified by Strnad et al. [112] and Tarkowská and coworkers [113] in *Populus* spp., which has been studied in plant tissue culture [114], is a hydroxylated analogue of BA with a hydroxyl group attached at its N6 side chain which results in the formation of O-glucoside metabolites that can be reversibly sequestrated in plants to produce active cytokinin forms when needed. The presence of these free bases of mT proclaims higher activity in cell division, cell growth, and differentiation than the other cytokinins [115]. Many reports indicate that mT and its derivatives are useful in improving shoot production and secondary metabolite production and/or reducing tissue culture-induced abnormalities in some plant species such as *Pyrus communis* [116–118]. The properties of mT subsequently enhance the rhizogenesis and acclimatization of plantlets [114]. mT has also been successfully employed in rooting of *Malus domestica* [119] and *Ananas comosus* [120], acclimatization in *Uniola paniculata* [121], and complete regeneration and acclimatization of *Aloe polyphylla* [122], *Sesamum indicum* [123], and *Salvia sclarea* [124]. Higher rooting percentage, root number, and length have been reported in *Prunus* rootstocks [125].

In vitro-propagated *C. sativa* plants using *mT* have been successfully hardened and grown to full maturity in soil without any detectable variation in morphological or growth characteristics. The regenerated plants are highly comparable with the mother plants in terms of physiological, biochemical, and genetic profile [60, 89].

2.3.1.4.3 Assessment of genetic stability of micropropagated plants

Molecular marker is an important tool being used in the analysis of genetic fidelity and true-to-type nature of micropropagated plants. Although clonal propagation through tissue culture should generate clones similar to the mother plant, the use of plant growth regulators, culture conditions such as light and humidity, and frequency of subculture can initiate genetic instability due to genetic and epigenetic (somaclonal) variations among the regenerated plants [126]. Morphological, cytological, and biochemical changes occurring in plants can be an indicator of genetic alterations. Hence, a quality checkup for true-to-type planting material at an early stage of development is considered to be very useful in plant tissue culture [127].

In comparison to various morphological, cytological, and protein markers used for detection of variations in tissue-cultured plants, molecular markers are stable, highly reproducible, detectible in almost all tissues, reliable, easy to access, and fast to assay, independent of development environment, and thus have been successfully employed to assess the genomic stability of regenerated plants.

In our laboratory, a DNA-based inter-simple sequence repeat (ISSR) markers have been successfully used to monitor the genetic stability of the micropropagated plants of *C. sativa*. ISSR technique is simple, fast, cost-effective, and a highly reliable technique in which a single sequence repeat motif is used as a primer for amplification of regions between microsatellites. The regenerated cannabis plants were stable up to 30 passages in culture and hardened in soil for 8 months [128]. Fifteen ISSR primers generated a total of 115 distinct and reproducible bands. The banding pattern for each primer was uniform and comparable to the mother plant from which the cultures had been established. Based on our results, the true-to-type nature of the in vitro raised clones was confirmed using ISSR markers.

The ISSR molecular markers have also been used in the evaluation of the genetic stability of different micropropagated plant species, such as *Glycyrrhiza glabra* L. [129] and *Tetrastigma hemsleyanum* [130]. Based on our results, the micropropagation protocol standardized for *C. sativa* can be used commercially with a minimum possibility of any in vitro-induced variability.

2.3.1.4.4 Assessment of chemical profile and cannabinoid content of micropropagated plants

GC-FID has been used to assess cannabinoid profile and content to identify differences, if any, in the chemical constituents of in vitro propagated plants versus vegetatively grown plants and mother plant. In general, THC content in all groups increased with plant age up to a highest level during budding stage, where the THC content reached a plateau before the onset of senescence. The pattern of changes observed in the concentration of other cannabinoid content with plants age has followed a similar trend in all groups of plants. Qualitatively, the cannabinoid profile obtained using GC-FID, in mother plants, vegetatively grown plants, and in vitro propagated plants, was found to be similar to each other and to that of the mother plant of *C. sativa*. Minor differences were observed in cannabinoid concentration within and among the groups were not statistically significant [128, 131]. These results confirm the clonal fidelity of in vitro propagated plants of *C. sativa* and suggest that the biochemical mechanism used in this study to produce the micropropagated plants does not affect the metabolic content and can be used for mass propagation of true-to-type plants of this species.

2.3.1.4.5 Conservation of cannabis germplasm

The conservation of germplasm can take advantage of innovative techniques which allow preservation in vitro (slow growth storage) or in liquid nitrogen (cryopreservation) of the plant material. Slow growth storage refers to techniques enabling the in vitro conservation of shoot cultures in aseptic conditions by reducing markedly the frequency of periodic subculturing, without affecting the viability and regrowth of shoot cultures. Cryopreservation refers to the storage of explants from tissue culture at ultra-low temperature (−196 °C).

In order to develop a secure and stable in vitro clonal repository of the elite clones of cannabis, germplasm conservation of a high Δ^9-THC yielding a variety of *C. sativa* L. was successfully attempted in our laboratory by using synthetic seed technology and media supplemented with osmotic agents [100], cryopreservation using axillary buds [102], and shoot tips [132]. Explants of nodal segments containing single axillary bud were excised from proliferated shoot cultures and encapsulated in high-density sodium alginate (230 mM) hardened by 50 mM $CaCl_2$. The "encapsulated" (synthetic seeds) nodal segments were stored at 5, 15, and 25 °C for 8, 16, and 24 weeks and monitored for the regrowth and survival frequency under the tissue culture conditions (16 h photoperiod, 25 °C) on MS medium supplemented with TDZ (0.5 µM). "Encapsulated" nodal segments could be stored at low temperature of 15 °C up to 24 weeks with maximum regrowth ability and survival frequency of 60%. Well-developed plantlets regenerated from "encapsulated" nodal segments were successfully acclimatized inside the grow room with 90% survival frequency.

The protocols developed would be helpful for large-scale mass propagation of elite cannabis varieties for further use in phytopharmaceuticals.

2.3.2 Outdoor cultivation

Cannabis cultivated outdoors need full sunlight to grow profusely. The outdoor cultivation of cannabis in Mississippi (Figure 2.13) starts at late March/early April, when the weather starts warming up, and could last till November or early December depending on the variety.

Starting from seeds, plants may be raised in small biodegradable (jiffy) pots and well-rooted healthy seedlings may than be planted in the field. Alternatively, seeds can also be directly planted to the field. Male flowers start appearing within 2–3 months (around middle of July) followed by female flowers. Male plants are generally removed from fields for several reasons: (1) male plants contain less THC as compared to female plants; (2) to avoid pollination within a variety which produces seeds in mature crop and results in less yield of biomass and ultimately THC as compared to the seedless (sinsemilla) mature plants; (3) to avoid cross-pollination (if different varieties are grown in close plots) among the varieties. It is difficult to maintain consistency in chemical profiles of selected high THC-producing genotypes under field conditions if grown from seeds. Therefore, vegetatively propagated cuttings of a screened and selected mother plants (based on its chemical profile) are used to cultivate biomass batches consistent in its chemical profile. Similar to propagation from seed, cuttings can be raised in biodegradable jiffy pots, and well-rooted cuttings are planted directly in field using automated planter.

Tetrahydrocannabinol (THC) content increases with the age of plant, reaching the highest level at the budding stage and achieve a plateau before the onset of senescence. The maturity of the crop is determined visually and confirmed based on the

Figure 2.13: Outdoor cultivation of *Cannabis sativa* L.: (A) crop at vegetative stage and (B) flowering stage.

THC and other cannabinoid content (using GC-FID) in samples collected at different growth stages of the plants. Since the whole plant does not mature at the same time, mature upper buds are harvested first and other branches are given more time to achieve their maturity. Field-cultivated cannabis plants are generally bigger and contain higher biomass compared to indoor grown plants.

Other than field plantation, cuttings can be grown in hydroponic systems outdoors (Figure 2.10). Hydroponic cultivation is less labor intensive and produces a cleaner harvest as compared to cultivation in soil.

2.3.2.1 Harvesting

Identifying the optimum harvesting stage is a critical step in cannabis cultivation. Daily monitoring of the THC content allows the harvesting material with the desired

THC content. Since it is observed that the levels of THC are higher during the morning hours and gradually decrease with the noon and afternoon hours, harvesting is recommended during the morning hours. Within the plant, the top mature buds may be harvested first and the rest of the immature buds may allow some time to mature. Figure 2.13B shows a field-grown mature, ready-to-harvest plants of *C. sativa* L.

2.3.2.2 Postharvest processing

Hygiene of biomass material should be the utmost priority during harvest. If the biomass is being used as a starting material for pharmaceutical interest, its contact with the ground should be avoided. Dry and large leaves may be removed from mature buds before drying.

Depending on the size of cultivation, drying facility can be selected. For large-scale cultivation, the plants are dried in industrial-grade "forced-air" drying barn, such as BulkTobac, Gas-Fired Products, Inc., Charlotte, NC, USA, and for the small samples, simple laboratory oven may be used for overnight drying at 40 °C.

Once the plant material is dried properly, it can be hand manicured. Big leaves, left from the clipping before drying, should be separated from buds. These buds can be gently rubbed through screens of different sizes to separate small stems and seeds (if any) from the dried biomass. Automated machines designed for plant processing can also be used to separate big stems and seeds from the useable biomass.

Dried and processed cannabis biomass can be stored in an FDA-approved sealed fiber drums containing polyethylene bags inside at ~18–20 °C for short-term storage. However, for long-term storage, −10 °C (freezer), in the absence of light, is recommended. Stability of Δ^9-THC and other cannabinoid content in cannabis biomass and its products is reviewed by several authors [133–135]. Extraction of plant materials is done either by supercritical fluid extraction or solvent extraction. Decarboxylation of acid cannabinoids to the neutral cannabinoids is accomplished using the extract, where the plant material itself is subjected to the decarboxylation step before extraction.

2.4 Extraction of cannabinoids

Extraction of plant material is accomplished by one of two different methods: (1) solvent extraction or (2) supercritical fluid extraction. Cannabis extract can be kept at −20 °C for long-term storage. Decarboxylation of the acid cannabinoids to the neutral cannabinoids can be done either before extraction by heating the biomass or after extraction by heating the extract.

2.5 Current status and future prospects: cannabis-based drug development

Plant-based natural products play an important role in modern medicine. The process, however, is costly and time consuming. Therefore, using traditional knowledge with scientific focus would lead to developing botanical-based drugs to treat a specific disease condition.

In the United States, the Food and Drug Administration has developed guidance for the development of "Botanical Drugs" Products [136]. In the case of cannabis, Sativex® (THC and CBD in a 1:1 ratio) is the only pharmaceutical product that contains phytocannabinoids (GW Pharmaceuticals), which is available mainly in European and Canadian market. The prospect for developing new cannabis-based pharmaceuticals is wide open, and the change in the regulatory landscape in the United States will certainly encourage that prospect.

References

[1]   Jiang HE, Li X, Zhao YX, Ferguson DK, Hueber F, Bera S, Wang YF, Zhao LC, Liu CJ, Li CS. A new insight into *Cannabis sativa* (Cannabaceae) utilization from 2500-year-old Yanghai Tombs, Xinjiang, China. Journal of Ethnopharmacology 2006; 108: 414–422.
[2]   Doyle E, Spence AA. Cannabis as a medicine? British Journal of Anesthesia 1995; 74: 359–361.
[3]   Zuardi AW. History of Cannabis as a medicine: A review. Brazilian Journal of Psychiatry 2006; 28: 153–157.
[4]   Guindon J, Hohmann AG. The endocannabinoid system and pain. CNS & Neurological Disorders-Drug Targets (Formerly Current Drug Targets-CNS & Neurological Disorders) 2009; 8(6): 403–421.
[5]   Järvinen T, Pate DW, Laine K. Cannabinoids in the treatment of glaucoma. Pharmacology and Therapeutics 2002; 95(2): 203–220.
[6]   Slatkin NE, et al. Cannabinoids in the treatment of chemotherapy-induced nausea and vomiting: Beyond prevention of acute emesis. Journal of Supportive Oncology 2007; 5: 1–9.
[7]   Viveros MP, Marco EM, Llorente R, Lopez-Gallardo M. Endocannabinoid system and synaptic plasticity: Implications for emotional responses. Neural Plasticity 2007: 1–12.
[8]   Liang YC, Huang CC, Hsu KS. Therapeutic potential of cannabinoids in trigeminal neuralgia. Current Drug Targets-CNS & Neurological Disorders 2004; 3(6): 507–514.
[9]   Abrams DI, Jay CA, Shade SB, Vizoso H, Reda H, Press S, Kelly ME, Rowbotham MC, Petersen KL. Cannabis in painful HIV-associated sensory neuropathy: A randomized placebo-controlled trial. Neurology 2007; 68: 515–521.
[10] Pryce G, Baker D, et al. Emerging properties of cannabinoid medicines in management of multiple sclerosis. Trends in Neuroscience 2005; 28: 272–276.
[11] Tzadok M, Uliel-Siboni S, Linder I, Kramer U, Epstein O, Menascu S, Nissenkorn A, Yosef OB, Hyman E, Granot D, Dor M. CBD-enriched medical cannabis for intractable pediatric epilepsy: The current Israeli experience. Seizure 2016; 35: 41–44.
[12] Kriese U, Schumann E, Weber WE, Beyer M, Brhl L, Matthus B. Oil content, tocopherol composition and fatty acid patterns of the seeds of 51 *Cannabis sativa* L. genotypes. Euphytica 2004; 137: 339–351.

[13]   Nelson RA. (1996) Hemp and history, Rex Research, Jean NV. Accessed on the web Sep, 04 20016, http://www.rexresearch.com/hhist/hhicon.htm.
[14]   Schultes RE. Random thoughts and queries on the botany of cannabis. J. & A. Churchill, London, 1970; 11–33.
[15]   Small E, Marcus D. Hemp: A new crop with new uses for North America. Trends in New Crops and New Uses 2002; 24(5): 284–326.
[16]   Turner CE, ElSohly MA, Boeren EG. Constituents of *C. sativa* L. XVII. A review of the natural constituents. Journal of Natural Products 1980; 43: 169–234.
[17]   Wood TB, Spivey WTN, Easterfield TH. III-Cannabinol. Part I. Journal of the Chemical Society, Transactions 1899; 75: 20–36.
[18]   Adams R, Baker BR, Wearn RB. Structure of cannabinol. III. Synthesis of cannabinol, 1-hydroxy-3-amyl-6,6,9-trimethyl-6-dibenzopyran. Journal of the American Chemical Society 1940a; 62: 2204–2207.
[19]   Cahn RS. *Cannabis indica* resin. III. Constitution of cannabinol. Journal of the Chemical Society 1932; 3: 1342–1353.
[20]   Ghosh R, Todd AR, Wilkinson S. *Cannabis indica*. Part IV. The synthesis of some tetrahydrodibenzopyran derivatives. Journal of the American Chemical Society 1940; 1121–1125.
[21]   Adams R, Hunt M, Clark JH. Structure of cannabidiol, a product isolated from the marihuana extract of Minnesota wild hemp. Journal of the American Chemical Society 1940b; 62: 196–200.
[22]   Mechoulam R, Shvo Y. The structure of cannabidiol. Tetrahedron 1963; 19: 2073–2078.
[23]   Gaoni Y, Mechoulam R. Isolation, structure and partial synthesis of an active constituent of hashish. Journal of the American Chemical Society 1964; 86: 1646.
[24]   ElSohly MA, Radwan MM, Gul W, Chandra S, Galal A. Phytochemistry of *Cannabis sativa* L. Phytocannabinoids 2017; 103: 1–36.
[25]   Radwan MM, Chandra S, Gul S, ElSohly MA. Cannabinoids, phenolics, terpenes and alkaloids of cannabis. Molecules 2021; 26(9): 2774.
[26]   Grinspoon L, Bakalar JB. Marihuana, the forbidden medicine. New Haven, Yale University Press, 1993.
[27]   Mattes RD, Egelman K, Shaw LM, ElSohly MA. Cannabinoids appetite stimulation. Pharmacology, Biochemistry and Behavior 1994; 49(1): 187.
[28]   Brenneisen R, Egli A, ElSohly MA, Henn V, Spiess Y. The effect of orally and rectally administered Δ^9-tetrahydrocannabinol on spasticity: A pilot study with 2 patients. International Journal of Clinical Pharmacology and Therapeutics 1996; 34: 446–452.
[29]   Devinsky O, Cilio MR, Cross H, Fernandez-Ruiz J, French J, Hill C, Katz R, Di Marzo V, Jutras-Aswad D, Notcutt WG, Martinez-Orgado J. Cannabidiol: Pharmacology and potential therapeutic role in epilepsy and other neuropsychiatric disorders. Epilepsia 2014; 55(6): 791–802.
[30]   Small E, Marcus D. Tetrahydrocannabinol levels in hemp (*Cannabis sativa*) germplasm resources. Economic Botany 2003; 57: 545–558.
[31]   Shoyama Y, Hirano H, Oda M, Somehara T, Nishioka I. Cannabis. IX. Cannabichromevarin and cannabigerovarin, two new propyl homologs of cannabichromene and cannabigerol. Chemical and Pharmaceutical Bulletin 1975; 23: 1894–1895.
[32]   Kajima M, Piraux M. The biogenesis of cannabinoids in *Cannabis sativa*. Phytochemistry 1982; 21: 67–69.
[33]   Fellermeier M, Zenk MH. Prenylation of olivetolate by a hemp transferase yields cannabigerolic acid, the precursor of tetrahydrocannabinol. FEBS Letters 1998; 427: 283–28.
[34]   Sirikantaramas S, Taura F, Tanaka Y, Ishikawa Y, Morimoto S, Shoyama Y. Tetrahydrocannabinolic acid synthase, the enzyme controlling marijuana psychoactivity is secreted into the storage cavity of the glandular trichomes. Plant and Cell Physiology 2005; 469: 1578–1582.

[35] Sirikantaramas S, Taura F, Morimoto S, Shoyama Y. Recent advances in *Cannabis sativa* research: Biosynthetic studies and its potential in biotechnology. Current Pharmaceutical Biotechnology 1998; 8(4): 237–243.

[36] Flores-Sanchez IJ, Verpoorte R. Secondary metabolism in cannabis. Phytochemistry Reviews 2008; 7: 615–639.

[37] Linnaeus C. Species plantarum. Stockholm, Laurentius Salvius, 1737; 1753.

[38] Lamark JB. Encyclope'die me'todique. Botanique: Paris-Liege 1785: 1783–1803.

[39] Small E, Cronquist A. A practical and natural taxonomy for Cannabis. Taxon 1976; 25: 405–435.

[40] Hillig KW. Genetic evidence for speciation in Cannabis (Cannabaceae). Genetic Resources and Crop Evolution 2005; 52: 161–180.

[41] Emboden WA. Cannabis, A polytypic genus. Economic Botany 1974; 28: 304–310.

[42] Hillig KW. A chemotaxonomic analysis of terpenoid variation in Cannabis. Biochemical Systematics and Ecology 2004; 32: 875–891.

[43] Hillig KW, Mahlberg PG. A chemotaxonomic analysis of cannabinoids variation in Cannabis (Cannabaceae). American Journal of Botany 2004; 91: 966–975.

[44] Small E. Morphological variation of Cannabis. Canadian Journal of Botany 2004; 53(10): 978–987.

[45] Small E. American law and the species problem in Cannabis. Science and semantics. Bulletin on Narcotics 1975b; 27: 1–20.

[46] Gilmore S, Peakall R, Robertson J. Short tandem repeat (STR) DNA markers are hypervariable and informative in *Cannabis sativa*: Implications for forensic investigations. Forensic Science International 1975b; 131: 65–74.

[47] Small E. Classification of *Cannabis sativa* L. in Relation to agricultural, biotechnological, medical and recreational utilization. In: Chandra S, Lata H, ElSohly MA (Eds.) *Cannabis Sativa* L.: Botany and biotechnology. Springer Publication, Springer cham 2017; 1–62.

[48] Piomelli D, Russo EB. The *Cannabis sativa* versus *Cannabis indica* debate: An interview with Ethan Russo, MD. Cannabis and Cannabinoid Research 2016 Jan 1; 1 (1): 44–46.

[49] Evert RF. Esau's plant anatomy: Meristems, cells, and tissues of the plant body: Their structure, function, and development. Hoboken, New Jersey, John Wiley & Sons, Inc, 2006.

[50] Hayward HE. The structure of economic plants. New York, Macmillan Co., 1938.

[51] ElSohly MA, Slade D. Chemical constituents of marijuana: The complex mixture of natural cannabinoids. Life Science 1938; 78: 539–548.

[52] Mandolino G, Bagatta M, Carboni A, Ranalli P, Meijer ED. Qualitative and quantitative aspects of the inheritance of chemical phenotype in Cannabis. Journal of Industrial Hemp 2003; 8: 51–72.

[53] Fetterman PS, Keith ES, Waller CW, Guerrero O, Doorenbos NJ, Quimby MW. Mississippi-grown *Cannabis sativa* L.: Preliminary observation on chemical definition of phenotype and variations in tetrahydrocannabinol content versus age, sex, and plant part. Journal of Pharmaceutical Sciences 1971; 60(8): 1246–1249.

[54] Small E, Beckstead HD. Cannabinoid phenotypes in *Cannabis sativa*. Nature 1973; 245: 147–148.

[55] Fournier G, Richez-Dumanois C, Duvezin J, Mathieu JP, Paris M. Identification of a new chemotype in *Cannabis sativa*: Cannabigerol-dominant plants, biogenetic and agronomic prospects. Planta Medica 1987; 53(3): 277–280.

[56] De Backer B, Debrus B, Lebrun P, Theunis L, Dubois N, Decock L, Verstraete A, Hubert P, Charlier C. Innovative development and validation of an HPLC/DAD method for the qualitative and quantitative determination of major cannabinoids in cannabis plant material. Journal of Chromatography B 2009; 877(32): 4115–4124.

[57] Bócsa I, Máthé P, Hangyel L. Effect of nitrogen on tetrahydrocannabinol (THC) content in hemp (*Cannabis sativa* L.) leaves at different positions. Journal of the International Hemp Association 1997; 4(2): 78–79.

[58] De Meijer EP, Bagatta M, Carboni A, Crucitti P, Moliterni VC, Ranalli P, Mandolino G. The inheritance of chemical phenotype in *Cannabis sativa* L. Genetics 2003; 163(1): 335–346.

[59] Chandra S, Lata H, Khan IA, ElSohly MA. Photosynthetic response of *Cannabis sativa* L. to variations in photosynthetic photon flux densities, temperature and CO_2 conditions. Physiology and Molecular Biology of Plants 2008; 14: 299–306.

[60] Chandra S, Lata H, Mehmedic Z, Khan IA, ElSohly MA. Assessment of cannabinoids content in micropropagated plants of *Cannabis sativa* L. and their comparison with conventionally propagated plants and mother plant during developmental stages of growth. Planta Medica 2010; 76: 743–750.

[61] de Meijer ED, Van der Kamp HJ, Van Eeuwijk FA. Characterisation of Cannabis accessions with regard to cannabinoid content in relation to other plant characters. Euphytica 1992; 62(3): 187–200.

[62] Hemphill JK, Turner JC, Mahlberg PG. Cannabinoid content of individual plant organs from different geographical strains of *Cannabis sativa* L. Journal of Natural Products. 1980; 43(1): 112–122.

[63] Mendoza I, Zamora R, Castro J. A seeding experiment for testing tree-community recruitment under variable environments: Implications for forest regeneration and conservation in Mediterranean habitats. Biological Conservation 2009; 142(7): 1491–1499.

[64] Valle RS, King M. Existential-phenomenological alternatives for psychology. Oxford University Press, New York, 1978.

[65] Datwyler SL, Weiblen GD. Genetic variation in hemp and marijuana (*Cannabis sativa* L.) according to amplified fragment length polymorphisms. Journal of Forensic Sciences 2006;51(2):371–375.

[66] Techen N, Chandra S, Lata H, ElSohly MA, Khan IA. Genetic identification of female *Cannabis sativa* L. plants at early developmental stage. Planta Medica 2010; 16: 1938–1939.

[67] Devlin RM. Plant physiology, 3rd ed. Van Nortrand Company, New York, NY, 1975; 600 p.

[68] Sage RF, Sharkey TD. The effect of temperature on the occurrence of O_2 and CO_2 insensitive photosynthesis in field grown plants. Plant Physiology 1987; 84(3): 658–664.

[69] Borjigidai A, Hikosaka K, Hirose T, Hasegawa T, Okada M, Kobayashi K. Seasonal changes in temperature dependence of photosynthetic rate in rice under a free-air CO_2 enrichment. Annals of Botany 2006; 97(4): 549–557.

[70] Hikosaka K, Ishikawa K, Borjigidai A, Muller O, Onoda Y. Temperature acclimation of photosynthesis: Mechanisms involved in the changes in temperature dependence of photosynthetic rate. Journal of Experimental Botany 2006; 57(2): 291–302.

[71] Nagai T, Makino A. Differences between rice and wheat in temperature responses of photosynthesis and plant growth. Plant and Cell Physiology 2009; 50(4): 744–755.

[72] Chandra S, Lata H, Khan IA, ElSohly MA. Temperature response of photosynthesis in different drug and fiber varieties of *Cannabis sativa* L. Physiology and Molecular Biology of Plants 2012a; 17(3): 297–303.

[73] Kimball BA, Idso SB. Increasing atmospheric CO2: Effects on crop yield, water use and climate. Agricultural Water Management 1983; 7(1–3): 55–72.

[74] Cure JD. Carbon dioxide doubling responses: A crop survey. Direct Effects of Increasing Carbon Dioxide on Vegetation 1985 Dec: 99–116.

[75] Cure JD, Acock B. Crop responses to carbon dioxide doubling: A literature survey. Agricultural and Forest Meteorology 1986; 38(1–3): 127–145.

[76] Ceulemans R, Van Praet L, Jiang XN. Effects of CO_2 enrichment, leaf position and clone on stomatal index and epidermal cell density in poplar. New Phytologist 1995; 131(1): 99–107.

[77] Idso KE, Idso SB. Plant responses to atmospheric CO_2 enrichment in the face of environmental constraints: A review of the past 10 years' research. Agricultural and Forest Meteorology 1994; 69 (3–4): 153–203.

[78] Zelitch I. Improving the efficiency of photosynthesis: The opportunity exists to increase crop productivity by regulating wasteful respiratory processes. Science 1975; 188(4188): 626–633.

[79]   Poorter H. Interspecific variation in the growth response of plants to an elevated ambient CO_2 concentration. In: Rozema J, Lambers H, Van de Geijn SC, Cambridge ML (eds) CO_2 and Biosphere. Advances in vegetation science, Vol 14. Springer, Dordrecht, 1993; 77–98.

[80]   Chandra S, Lata H, Khan IA, ElSohly MA. Photosynthetic response of *Cannabis sativa* L., an important medicinal plant, to elevated levels of CO2. Physiology and Molecular Biology of Plants 2012b; 17(3): 291–295.

[81]   Fisse J, Braut F, Cosson L, Paris M. In vitro study of the organogenetic capacities of *Cannabis sativa* L. tissues: Effect of different growth substances auxins, cytokinins, gibberellins. Plantes Medicinales Et Phytotherapie 1981; 15(4): 217–223.

[82]   Hartsel SC, Loh WT, Robertson LW. Biotransformation of cannabidiol to cannabielsoin by suspension cultures of *Cannabis sativa* and *Saccharum officinarum*. Planta Medica 1983; 48(5): 17–19.

[83]   Braemer R, Paris M. Biotransformation of cannabinoids by a cell suspension culture of *Cannabis sativa* L. Plant Cell Reports 1987; 6(2): 150–152.

[84]   Ranalli P, Candilo MD, Mandolino G, Grassi G, Carboni A. Hemp for sustainable agricultural systems. Agro Industry 1999; 10: 33–38.

[85]   Feeney M, Punja ZK. Tissue culture and Agrobacterium-mediated transformation of hemp (*Cannabis sativa* L.). In Vitro Cellular & Developmental Biology-Plant 2003; 39(6): 578–585.

[86]   Slusarkiewicz-Jarzina AU, Ponitka A, Kaczmarek Z. Influence of cultivar, explant source and plant growth regulator on callus induction and plant regeneration of *Cannabis sativa* L. Acta Biologica Cracoviensia Series Botanica 2005; 47: 145–151.

[87]   Bing X, Ning L, Jinfeng T, Nan G. Rapid tissue culture method of *Cannabis sativa* for industrial uses.CN 1887043 A 20070103 Patent, 2007; p 9.

[88]   Lata H, Chandra S, Khan IA, ElSohly MA. Thidiazuron induced high frequency direct shoot organogenesis of *Cannabis sativa* L. In-vitro Cellular and Developmental Biology-Plant 2009a; 45: 12–19.

[89]   Lata H, Chandra S, Khan I, ElSohly MA, ElSohly MA. Propagation through alginate encapsulation of axillary buds of *Cannabis sativa* L.– An important medicinal plant. Physiology and Molecular Biology of Plants 2009b; 15(1): 79–86.

[90]   Tang CH, Wang XS, Yang XQ. Enzymatic hydrolysis of hemp (*Cannabis sativa* L.) protein isolate by various proteases and antioxidant properties of the resulting hydrolysates. Food Chemistry. 2009;114(4):1484–1490.

[91]   Lata H, Chandra S, Khan IA, ElSohly MA. High frequency plant regeneration from leaf derived callus of high Δ9-tetrahydrocannabinol yielding *Cannabis sativa* L. Planta Medica 2010a; 76: 1629–1633.

[92]   Lata H, Chandra S, Techen N, Khan IA, ElSohly MA. Assessment of genetic stability of micropropagated plants of *Cannabis sativa* L. by ISSR markers. Planta Medica 2010b; 76: 97–100.

[93]   Condic ML. Totipotency: What it is and what it is not. Stem Cells and Development 2014; 23(8): 796–812.

[94]   Sharma S, Kashyap S, Vasudevan P. Development of clones and somaclones involving tissue culture, mycorrhiza and synthetic seed technology. Journal of Scientific and Industrial Research 2000; 59(7): 531–540.

[95]   Beard KM, Boling AW, Bargmann BO. Protoplast isolation, transient transformation, and flow-cytometric analysis of reporter-gene activation in Cannabis sativa L. Industrial Crops and Products 2021; 164: 113–360.

[96]   Adams TK, Masondo NA, Malatsi P, Makunga NP. *Cannabis sativa*: From therapeutic uses to micropropagation and beyond. Plants 2021; 10(10): 2078.

[97]   Delporte F, Pretova A, Du Jardin P, Watillon B. Morpho-histology and genotype dependence of *in vitro* morphogenesis in mature embryo cultures of wheat. Protoplasma 2014; 251(6): 1455–1470.

[98]   Valladares S, Varas E, Vielba JM, Vidal N, Codesido V, Castro R, Sanchez C. Expression of a Rap2. 12 like-1 ERF gene during adventitious rooting of chestnut and oak microshoots. Israel Journal of Plant Sciences. 2020; 67 (1–2): 69–82.

[99]   Lata H, Chandra S, Techen N, Khan IA, ElSohly MA. *In vitro* mass propagation of *Cannabis sativa*: A protocol refinement using novel aromatic cytokinin *meta*-topolin and the assessment of eco-physiological, biochemical and genetic fidelity of micropropagated plants. Journal of Applied Research on Medicinal and Aromatic Plants 2016; 3: 18–26.

[100]   Lata H, Chandra S, Techen N, Khan IA, ElSohly MA. *In vitro* germplasm conservation of high Δ9-tetrahydrocannabinol yielding elite clones of *Cannabis sativa* L. under slow growth conditions. Acta Physiologiae Plantarum 2012; 34(2): 743–750.

[101]   Lata H, Chandra S, Techen N, Wang YH, Khan IA. Molecular analysis of genetic fidelity in micropropagated plants of *Stevia rebaudiana* Bert. using ISSR markers. American Journal of Plant Sciences 2013; 4: 964–971.

[102]   Lata H, Uchendu E, Chandra S, Majumdar CG, Khan IA, ElSohly MA. Cryopreservation of axillary buds of Cannabis sativa L. by V-cryoplate droplet-vitrification: The critical role of sucrose preculture. Cryo Letters 2019; 40(5): 291–298.

[103]   Torrey JG. The initiation of organized development in plants. Advances in Morphogenesis 1966; 5: 39–91.

[104]   Lata H, Bedir E, Hosick A, Ganzera M, Khan I, Moraes RM. In vitro plant regeneration from leaf-derived callus in Black cohosh (*Cimicifuga racemosa*). Planta Medica 2002; 68: 912–915.

[105]   Faisal M, Anis M. Rapid mass propagation of *Tylophora indica* Merrill via leaf callus culture. Plant Cell Tissue and Organ Culture 2003; 75(2): 125–129.

[106]   Laloue M, Pethe C. Dynamics of cytokinin metabolism in tobacco cells. In: Wareing PE (Ed.) Plant growth substances. London, UK, Academic, 1982; 185–195.

[107]   Bespalhok FJC, Hattori K. Friable embryogenic callus and somatic embryo formation from cotyledon explants of African marigold (*Tagetes erecta* L.). Plant Cell Reports 1998; 17(11): 870–875.

[108]   Saxena PK, Malik KA, Gill R. Induction by thidiazuron of somatic embryogenesis in intact seedlings of peanut. Planta 1992; 187(3): 421–424.

[109]   Dey M, Bakshi S, Galiba G, Sahoo L, Panda SK. Development of a genotype independent and transformation amenable regeneration system from shoot apex in rice (*Oryza sativa* sp. *indica*) using TDZ. Biotech 2012; 2(3): 233–240.

[110]   Thinh NT. Cryopreservation of germplasm of vegetatively propagated tropical monocots by vitrification (Doctoral dissertation).

[111]   Lata H, Chandra S, Techen N, Khan IA, ElSohly MA. In vitro mass propagation of *Cannabis sativa* L.: A protocol refinement using a novel aromatic cytokinin meta-topolin and assessment of eco-physiological, biochemical and genetic fidelity of micropropagated plants. Journal of Applied Research on Medicinal and Aromatic Plants 2016; 3(1): 18–26.

[112]   Strnad M, Hanuš J, Vaněk T, Kamínek M, Ballantine JA, Fussell B, Hanke DE. Meta-topolin, a highly active aromatic cytokinin from poplar leaves (Populus× canadensis Moench., cv. Robusta). Phytochemistry 1997; 45(2): 213–218.

[113]   Novák O, Tarkowski P, Tarkowská D, Doležal K, Lenobel R, Strnad M. Quantitative analysis of cytokinins in plants by liquid chromatography–single-quadrupole mass spectrometry. Analytica Chimica Acta 2003; 480(2): 207–218.

[114]   Aremu AO, Bairu MW, Szüčová L, Doležal K, Finnie JF, Van Staden J. Assessment of the role of meta-topolins on *in vitro* produced phenolics and acclimatization competence of micropropagated 'Williams' banana. Acta Physiologiae Plantarum 2012; 34(6): 2265–2273.

[115]   Werbrouck SP, Strnad M, Van Onckelen HA, Debergh PC. Meta-topolin, an alternative to benzyladenine in tissue culture. Physiologia Plantarum 1996; 98(2): 291–297.

[116]   Dimitrova N, Nacheva L, Berova M. Effect of meta-topolin on the shoot multiplication of pear rootstock OHF-333 (*Pyrus communis* L.). Acta Scientiarum Polonorum. Hortorum Cultus 2016; 15(2): 43–53.

[117] Khanam MN, Javed SB, Anis M, Alatar AA. meta-Topolin induced in vitro regeneration and metabolic profiling in Allamanda cathartica L Industrial Crops and Products 2020 Mar; 1(145): 111–944.

[118] Kucharska D, Orlikowska T, Maciorowski R, Kunka M, Wójcik D, Pluta S. Application of meta-Topolin for improving micropropagation of gooseberry (Ribes grossularia). Scientia Horticulturae 2020; 272: 109–529.

[119] Dobránszki J, Magyar-Tábori K, Tombácz E. Comparison of the rheological and diffusion properties of some gelling agents and blends and their effects on shoot multiplication. Plant Biotechnology Reports 2011; 5(4): 345–352.

[120] Teklehaymanot T, Wannakrairoj S, Pipattanawong N. Meta-topolin for pineapple shoot multiplication under three in vitro systems. American-Eurasian Journal of Agricultural and Environmental Science 2010; 7(2): 157–162.

[121] Valero-Aracama C, Kane ME, Wilson SB, Philman NL. Substitution of benzyladenine with meta-topolin during shoot multiplication increases acclimatization of difficult-and easy-to-acclimatize sea oats (*Uniola paniculata* L.) genotypes. Plant Growth Regulation 2010; 60(1): 43–49.

[122] Bairu MW, Stirk WA, Dolezal K, et al. Optimizing the micropropagation protocol for the endangered Aloe polyphylla: Can meta-topolin and its derivatives serve as replacement for benzyladenine and zeatin. Plant Cell, Tissue and Organ Culture 2007; 90: 15–23.

[123] Elayaraja D, Sathiyamurthi S. Influence of organic manures and micronutrients fertilization on the soil properties and yield of sesame (Sesamum indicum L.) in coastal saline soil Indian Journal of Agricultural Research 2020 Feb; 1(54): 89–94.

[124] Erişen S, Kurt-Gür G, Servi H. In vitro propagation of *Salvia sclarea* L. by meta-Topolin, and assessment of genetic stability and secondary metabolite profiling of micropropagated plants. Industrial Crops and Products 2020; 157: 112–892.

[125] Gentile A, Jàquez Gutiérrez M, Martinez J, et al. Effect of meta-Topolin on micropropagation and adventitious shoot regeneration in Prunus rootstocks. Plant Cell, Tissue and Organ Culture (PCTOC) 2014; 118: 373–381.

[126] Vázquez AM, Linacero R. Stress and somaclonal variation. In: Pua, E., Davey, M.(eds) Plant Developmental Biology-Biotechnological Perspectives, Springer, Berlin, Heidelberg, 2010; 45–64.

[127] Zilberman D, Henikoff S. Genome-wide analysis of DNA methylation patterns. Development 2007; 134(22): 3959-65.

[128] Lata H, Chandra S, Techen N, Khan IA, ElSohly MA. Assessment of genetic stability of micropropagated plants of *Cannabis sativa* L. by ISSR markers. Planta Medica 2009c; 76: 97–100.

[129] Mehrotra S, Khwaja O, Kukreja AK, Rahman L. ISSR and RAPD based evaluation of genetic stability of encapsulated micro shoots of *Glycyrrhiza glabra* following 6 months of storage. Molecular Biotechnology 2012; 52(3): 262–268.

[130] Peng X, Zhang TT, Zhang J. Effect of subculture times on genetic fidelity, endogenous hormone level and pharmaceutical potential of *Tetrastigma hemsleyanum* callus. Plant Cell Tissue and Organ Culture 2015; 122(1): 67–77.

[131] Chandra S, Lata H, Khan IA, Mehmedic Z, ElSohly MA. Assessment of cannabinoids content in micropropagated plants of Cannabis sativa L. and their comparison with conventionally propagated plants and mother plant during developmental stages of growth. Planta Medica Advance Online Publication 2009b; 76: 743–750.

[132] Uchendu E, Lata H, Chandra S, Khan IA, ElSohly MA. Cryopreservation of shoot tips of elite cultivars of Cannabis sativa L. by droplet vitrification. Med Cannabis Cannabinoids 2019 Apr 4; 2(1): 29–34.

[133] Turner CE, Hadley KW, Fetterman PS, Doorenbos NJ, Quimby MW, Waller C. Constituents of *Cannabis sativa* L. IV: Stability of cannabinoids in stored plant material. Journal of Pharmaceutical Sciences 1973; 62(10): 1601–1605.

[134] Narayanaswami K, Golani HC, Bami HL, Dau RD. Stability of *Cannabis sativa* L. samples and their extracts, on prolonged storage in Delhi. Bulletin on Narcotics 1978; 30(4): 57–69.

[135] Mehmedic Z, Chandra S, Slade D, Denham H, Foster S, Patel AS, Ross SA, Khan IA, ElSohly MA. Potency trends of Δ^9-THC and other cannabinoids in confiscated Cannabis preparations from 1993–2008. Journal of Forensic Science 2010; 55(5): 209–217.

[136] Food and Drug Administration (FDA). Botanical Drug Development Guidance for Industry, https://www.fda.gov/downloads/drugs/guidancecomplianceregulatoryinformationguidances/ucm458484. 2006; pdf, downloaded on 03/ 09/2018.

Mahmoud A. ElSohly*, Shahbaz W. Gul and Waseem Gul

3 A comprehensive review of cannabis phytocannabinoids

Abstract: *Cannabis sativa* is one of the most extensively studied plants globally. It has been cultivated for thousands of years across the world for a variety of uses. While the seeds are of good nutritional value, fibers are used in textile products due to their durability. The plant has long been used in both recreational and medicinal purposes due to its psychotic and therapeutic effects. As such, it is highly important to understand the composition of the cannabis plant. This chapter provides a brief synopsis of the history, isolation, and structural elucidation for each of the 129 constituents classified as cannabinoid-type constituents.

Keywords: cannabis, cannabinoids, history, isolation, structural elucidation, biomass

3.1 Introduction

For over 6,000 years, *Cannabis sativa* L. (*C. sativa*) has been cultivated across the world. Cannabis has many different names; marijuana (above ground portion, USA), bhang (dried leaves, India), hashish, or hasheesh and many other preparations of the plant [1]. *C. sativa* is the most widely abused drug by millions of people throughout the world where one form or another of cannabis is consumed, and about 18% of Americans used it at least once in 2019 [2].

This annual herbaceous plant has been used as a valuable fiber source for the majority of known history, constituting various textile products given its durability, strength, and water resistance [3]. The oldest evidence for use of this plant was uncovered in 1974 in China, dating from approximately 4,000 BC [4].

Cannabis itself is a unique plant that has more than 100 variants or races. In the past some of these variants were regarded as different species because of their different morphology and or constituents. Today, it is regarded as one genus belonging to family Cannabaceae.

---

*Corresponding author: Mahmoud A. ElSohly,** National Center for Natural Products Research and Department of Pharmaceutics and Drug Delivery School of Pharmacy, University of Mississippi, MS 38677, USA, e-mail: melsohly@olemiss.edu, https://orcid.org/0000-0002-0019-2001
Shahbaz W. Gul, ElSohly Laboratories, Inc., 5 Industrial Park Drive, Oxford, MS 38655, USA; Sally McDonnell Barksdale Honors College, University of Mississippi, MS 38677, USA; School of Pharmacy, University of Mississippi, MS 38677, USA, Ayub Medical College, Mansehra Road, Abbottabad, KPK, Pakistan
Waseem Gul, ElSohly Laboratories, Inc., 5 Industrial Park Drive, Oxford, MS 38655, USA

https://doi.org/10.1515/9783110718362-003

In 1964, Gaoni and Mechoulam [5] identified Δ^9-THC as the primary psychoactive constituent of *C. sativa.*

The term "cannabinoids" was first coined by Mechoulam and Gaoni in 1967 [6], where the term describes a group of terpenophenols restricted in occurrence only to *C. sativa* and not to any other plant, sparking the first surge of research for the next three decades [7]. After the initial surge, the number of publications, as well as interest in cannabis, slowly declined until the early 1990s.

The most common and biologically significant are the cannabinoids with a C5 side chain in the form of an *n*-pentyl (or *n*-amyl) group. The most common of these is Δ^9-tetrahydrocannabinol (Δ^9-THC), cannabidiol (CBD), and cannabinol (CBN). So far, 129 known cannabinoids have been isolated and characterized [8].

The discovery and cloning of the cannabinoid receptors, in addition to the identification of anandamide, an endogenous cannabinoid by Dr. Mechoulam, sparked renewed interest in cannabis and cannabinoids [9].

On November 5, 1996, California became the first state to legalize medical marijuana, paving the path to 37 other states legalizing medical marijuana in the years to come. The interest in pharmacological, therapeutic, and psychoactive effects of cannabinoids continues today, with groundbreaking research dealing with different derivatives of cannabinoids as compared to the parent cannabinoids.

3.2 Hemp and marijuana are not the same

It appears that the physical and chemical variations among cannabis numerous races are triggered by environmental factors. Cannabis variants are often divided into fiber types which produce little psychoactive ingredients but have strong phloem fibers useful in making ropes, for example, some Canadian and Turkish plants and phenodrug variants or drug type which are rich in psychoactive compounds, for example, South African, Mexican, and Columbian variants.

On February 7, 2014, the Agricultural Improvement Act of 2014 (2014 Farm Bill) was signed into law by President Obama. Four years later, a revision of the 2014 Farm Bill, dubbed the 2018 Farm Bill, was passed on December 11, 2018. The 2018 Farm Bill included a distinct separation between "hemp" and "marijuana." Hemp is not a controlled substance and must meet two conditions: (a) the cannabis plants and extracts from hemp must contain less than 0.3% equivalent of Δ^9-THC and (b) must be produced under a state program approved by the USDA. Any cannabis material that fails either of these two criteria is classified as marijuana.

Marijuana, on the other hand, is a Schedule I Controlled Substance and is subject to DEA and state regulations. This distinction, as well as the legalization of marijuana for both medicinal and recreational use, has generated much controversy, leaving the future of cannabis growth and research uncertain.

3.3 Historical review of the chemical research on cannabis

Throughout the nineteenth century there was a great deal of chemical work on natural products, especially alkaloids, for example, morphine, cocaine, strychnine, and many others. However, most of the terpenoids were not isolated until the end of the century or even much later. The reason behind this disparity is that alkaloids are relatively easy to isolate whereas terpenes occur mostly as difficult to separate mixtures. The research for the active constituents in cannabis took place in 1821 with no clear results. This was followed in 1840 by preparing a volatile oil fraction from cannabis, an alcohol extract in 1848, and then in 1885 a resin was obtained by extracting alkalinized plant material in the hope that the active ingredient might be an alkaloid.

At the turn of the century (1896) attempts were made to separate the active component by fractional distillation of an ether extract. This resulted in the production of an active fraction. This fraction was highly impure however because as it turned out many cannabinoids boil at the same temperature. Later in 1898–1899, this volatile fraction was further purified by acetylation to give a crystalline fraction that upon hydrolysis gave a pure crystalline compound $C_{21}H_{26}O_2$ which was named CBN. Thus, CBN represents the first natural cannabinoid to be obtained in pure form.

The isolation of CBN had a component of luck to it as it occurs in cannabis in very small amounts, unless it is too old, and in this case, some or most of the active constituent (Δ^9-THC) is oxidized by atmospheric air and light, producing CBN. The structure of CBN was established largely by Cahn [10].

3.4 Cannabinoid-type constituents of *C. sativa*

Over 500 total constituents of *C. sativa* have been identified, of which 129 are cannabinoids (Table 3.1). Cannabigerolic acid (CBGA) is the precursor for all of the major cannabinoids. Three different synthase enzymes then convert CBGA to the three other major cannabinoid acids: Δ^9-tetrahydrocannabinolic acid (THCA) synthase produces Δ^9-THCA; cannabidiolic acid (CBDA) synthase produces CBDA; and cannabichromenic acid (CBCA) synthase produces CBCA.

Upon irradiation of CBCA with UV light, cannabicyclolic acid (CBLA) is formed. Reacting CBDA with epoxidase results in the production of cannabielsoic acid (CBEA). Decarboxylation of THCA, CBDA, CBCA, CBEA, and CBLA through nonenzymatic means, such as the use of heat, yields the neutral form of each cannabinoid, namely THC, CBD, cannabichromene (CBC), cannabielsoin (CBE), and cannabicyclol (CBL), respectively. Below are the cannabis phytocannabinoids in their major types and their chemical structures.

Table 3.1: Number of cannabinoids in each chemical class in *Cannabis sativa*.

Chemical class	Number of cannabinoids
CBG	16
Δ^9-THC	25
CBT	9
Δ^8-THC	5
CBN	11
CBD	13
CBND	2
CBE	5
CBC	9
CBL	3
Miscellaneous	31
Total	**129**

3.4.1 CBG-type cannabinoids

There is a total of 16 cannabinoids in this cannabinoid class. The structures for these cannabinoids can be seen in Figures 3.1–3.3.

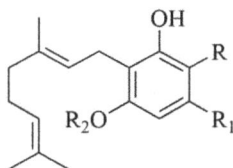

1 (E)-CBG,	R=H,	$R_1=C_5H_{11}$,	$R_2=H$
2 (E)-CBGM,	R=H,	$R_1=C_5H_{11}$,	$R_2=CH_3$
3 (E)-CBGA,	R=COOH,	$R_1=C_5H_{11}$,	$R_2=H$
4 (E)-CBGAM,	R=COOH,	$R_1=C_5H_{11}$,	$R_2=CH_3$
5 (E)-CBGV,	R=H,	$R_1=C_3H_7$,	$R_2=H$
6 (E)-CBGVA,	R=COOH,	$R_1=C_3H_7$,	$R_2=H$

Figure 3.1: Structures of six CBG-type cannabinoids (**1–6**).

Cannabigerol (*E*-CBG, **1**) was first isolated from cannabis resin in 1963. Florisil chromatography was used for the isolation of compound **1** and its structure was confirmed through synthesis [11]. Two years later, the monomethyl ether of *E*-CBG (CBGM, **2**) was isolated from a benzene extract of hemp. The extract was first heated in toluene for seven hours and then CBGM was purified through silica-gel column chromatography using benzene as the eluent [12]. In 1975, both CBGA (**3**) [13] and the monomethyl ether of CBGA (CBGAM, **4**) were isolated [14]. Cannabigerovarin (CBGV, **5**) and cannabigerovarinic acid (CBGVA, **6**) were both isolated from "Meao" variant of cannabis from Thailand. The chemical structures for both cannabinoids were determined using IR, NMR, and UV spectroscopy [15, 16]. The structures for these six CBG-type cannabinoids are shown in Figure 3.1.

The *cis* isomer of CBGA ((*Z*)-CBGA, **7**) was isolated through silica-gel column chromatography in 1995 from an acetone extract of the leaves of a Mexican strain of *C. sativa*. The chemical structure of (*Z*)-CBGA (**7**) was determined through NMR spectroscopy and confirmed via synthesis [17]. The polar dihydroxycannabigerol derivative (camagerol, **8**) was isolated from the aerial parts of the Carma *C. sativa* strain through reversed-phase (C$_{18}$) silica-gel column chromatography, followed by normal-phase silica gel chromatography, and then through normal phase-high performance liquid chromatography (NP-HPLC) in 2008. The wax of the aerial parts of the Carma strain was hydrolyzed and purified through silica gel and alumina column chromatography, which produced both waxy and nonwaxy fractions. From one of the waxy fractions, the farnesyl prenylogue of cannabigerol (sesquicannabigerol, **9**) was isolated [18]. The structure of sesquicannabigerol was elucidated by ^{1}H and ^{13}C-NMR techniques and semisynthesis. The structures of these three CBG-type cannabinoids are shown in Figure 3.2.

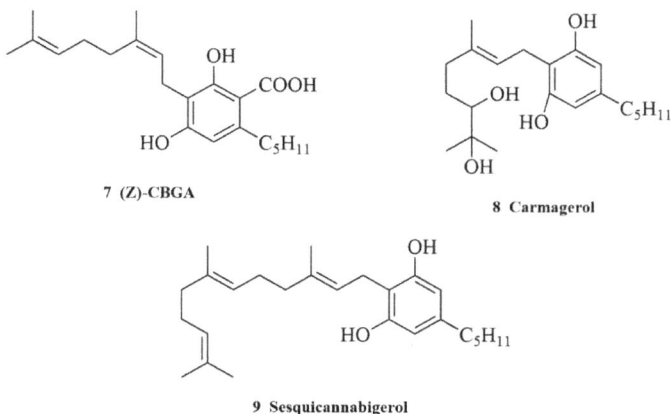

Figure 3.2: Structures of three CBG-type cannabinoids (**7–9**).

Four epoxy cannabigerol derivatives (**10–13**) were isolated from Mississippi grown high potency *C. sativa*. For the isolation and purification of these compounds, vacuum liquid chromatography (VLC), flash chromatography, and HPLC were used. The chemical structures of these derivatives were determined to be (±)-6,7-*cis*-epoxycannabigerol (**10**), (±)-6,7-*trans*-epoxycannabigerol (**11**), (±)-6,7-*cis*-epoxycannabigerolic acid (**12**), and (±)-6,

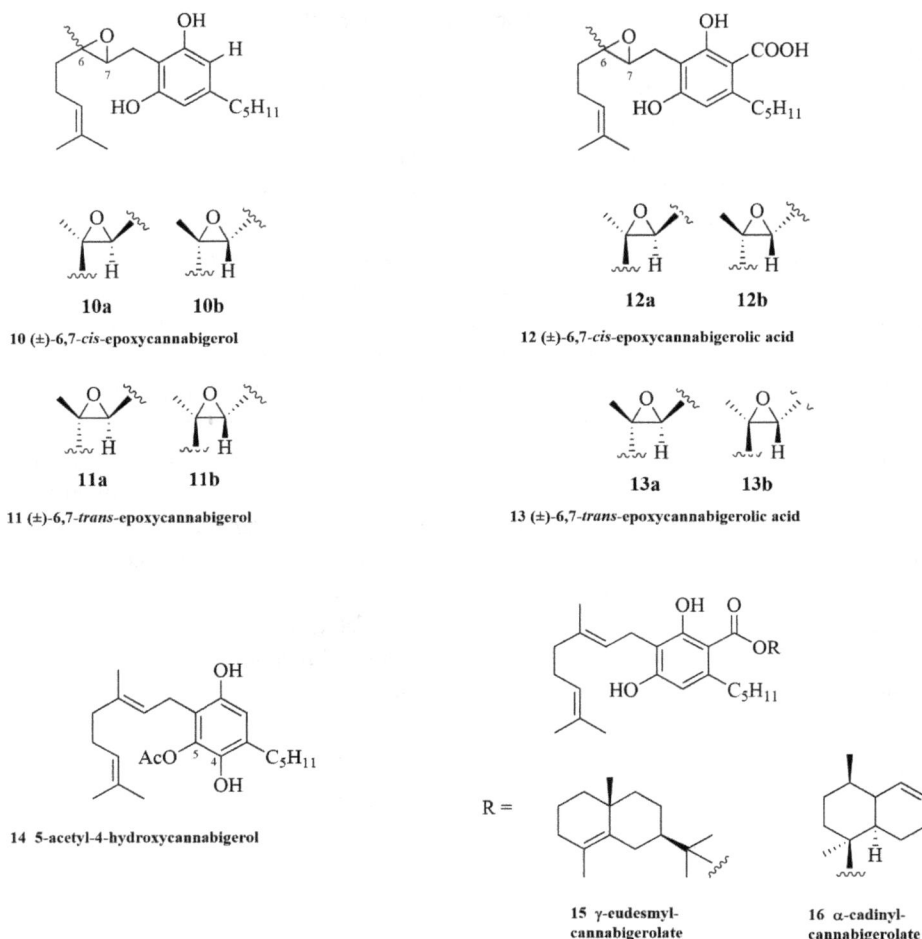

Figure 3.3: Structures of seven CBG-type cannabinoids (**10–16**).

7-*trans*-epoxycannabigerolic acid (**13**) using NMR and HRESIMS spectroscopic techniques [19]. The same group isolated 5-acetyl-4-hydroxy-cannabigerol (**14**) [20] from the buds of *C. sativa*, using NP-HPLC of polar fractions, as well as two esters of CBGA, namely γ-eudesmyl-cannabigerolate (**15**) and α-cadinyl-cannabigerolate (**16**), from high-potency *C. sativa* using chiral HPLC [21]. The structures of these seven CBG-type cannabinoids are shown in Figure 3.3.

When compared to the level of activity of Δ⁹-THC, the activity of CBG-type cannabinoids is negligible [22, 23]. CBD potentiated pentobarbital induction and sleeping time, while CBG antagonized it [24]. There is also relative data demonstrating the decrease in the rate of absorption and excretion of pentobarbital [25] due to CBG (**1**) as well as exhibiting activity against nuclear membrane bound ribosomes of infant rat brain cells [26]. Rat brain cortex slices were used to show the inhibiting effect of CBG (**1**) on the incorporation of both leucine and uridine into the proteins and nucleic acids of the slices, respectively [27].

3.4.2 Δ⁹-THC-type cannabinoids

There is a total of 25 cannabinoids in this class. The structures for these cannabinoids can be seen in Figures 3.4–3.6.

At least two different numbering systems are used in cannabinoids. The first system is mostly used in the USA, and it depends on pyran-type cannabinoids as dibenzopyrans as follows: 3-pentyl-6a,7,8,10a-tetrahydro-6,6,9-trimethyl-6 H-dibenzo[4]pyran-1-ol. The second system is used by Israeli and European scientists and it depends on cannabinoids as substituted p-mentha derivative (Δ¹-THC based on p-mentha). Using this system, the carbon atoms retain their numbers in going from pyran type to nonpyran-type cannabinoids. Furthermore, the system has a biogenetic significance as cannabinoids are monoterpene derivatives. The International Union of Pure and Applied Chemistry is yet to have a ruling on the system of numbers that should be used. The numbering systems can be seen in Figure 3.4.

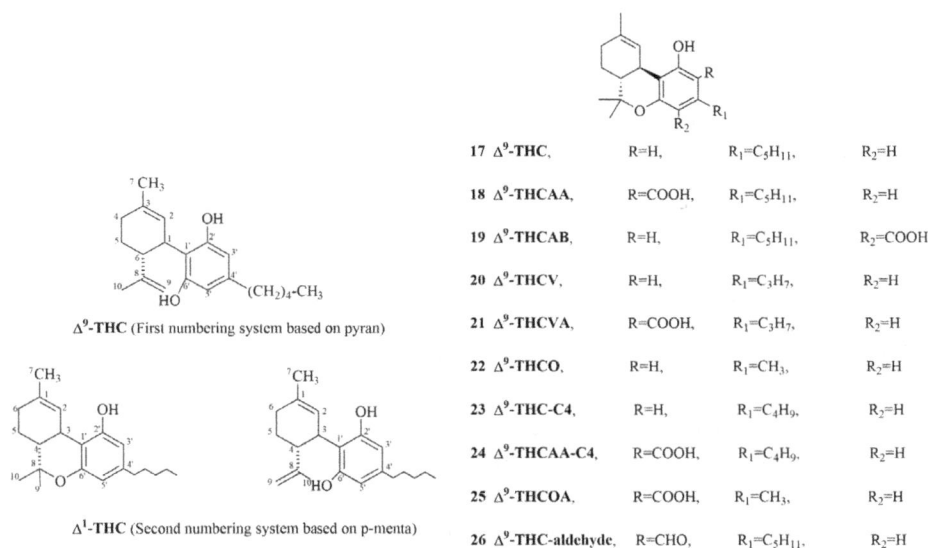

17 Δ⁹-THC,	R=H,	R_1=C_5H_{11},	R_2=H
18 Δ⁹-THCAA,	R=COOH,	R_1=C_5H_{11},	R_2=H
19 Δ⁹-THCAB,	R=H,	R_1=C_5H_{11},	R_2=COOH
20 Δ⁹-THCV,	R=H,	R_1=C_3H_7,	R_2=H
21 Δ⁹-THCVA,	R=COOH,	R_1=C_3H_7,	R_2=H
22 Δ⁹-THCO,	R=H,	R_1=CH_3,	R_2=H
23 Δ⁹-THC-C4,	R=H,	R_1=C_4H_9,	R_2=H
24 Δ⁹-THCAA-C4,	R=COOH,	R_1=C_4H_9,	R_2=H
25 Δ⁹-THCOA,	R=COOH,	R_1=CH_3,	R_2=H
26 Δ⁹-THC-aldehyde,	R=CHO,	R_1=C_5H_{11},	R_2=H

Δ⁹-THC (First numbering system based on pyran)

Δ¹-THC (Second numbering system based on p-menta)

Figure 3.4: Structures of ten Δ⁹-THC-type cannabinoids (**17–26**).

Pure (–)-*trans*-Δ⁹-THC (**17**), the predominant cannabinoid as well as the primary psychoactive ingredient of *C. sativa* was isolated in 1964 from a hexane extract of hashish using Florisil column chromatography [5]. Three years later, (–)-*trans*-Δ⁹-tetrahydrocannabinolic acid A (Δ⁹-THCAA, **18**) was isolated from cannabis using a cellulose powder column eluted with a mixture of hexane and dimethylformamide followed by preparative thin-layer chromatography (TLC) [28]. Another Δ^9-THC acid, namely (–)-*trans*-Δ^9-tetrahydrocannabinolic acid B (Δ^9-THCAB, **19**), was isolated from the purification of a hashish sole through a silicic acid column eluted with a mixture of ether in petroleum ether [29]. In 1971, (–)-*trans*-Δ^9-tetrahydrocannabivarin (Δ^9-THCV, **20**) was isolated from a light petroleum ether extract of a cannabis tincture of Pakistani origin using the countercurrent distribution technique [30].

Two years later, in 1973, (–)-*trans*-Δ^9-tetrahydrocannabivarinic acid (Δ^9-THCVA, **21**) was isolated from fresh *C. sativa* leaves from South Africa [31]. The same year, a new Δ^9-THC type cannabinoid, namely (–)-*trans*-Δ^9-tetrahydrocannabiorcol (Δ^9-THCO or Δ^9-THC-C_1, **22**) was identified in the light petroleum ether extract of Brazilian marijuana [32]. The ethyl acetate extracts of police confiscated cannabis leaves, resins, and tinctures were analyzed using gas chromatography coupled with mass spectrometry (GC–MS). From this analysis, new Δ^9-THC-type cannabinoids were identified and characterized, namely (–)-Δ^9-*trans*-THC-C_4 (**23**), (–)-Δ^9-*trans*-tetrahydrocannabinolic acid A-C_4 (Δ^9-THCAA-C_4, **24**), and (–)-Δ^9-*trans*-tetrahydrocannabiorcolic acid (Δ^9-THCOA, **25**) from the ethyl acetate extracts of police confiscated cannabis leaves, resins, and tinctures [33]. In 2015, the aldehyde derivative of Δ^9-THC, (–)-Δ^9-*trans*-tetrahydrocannabinal (Δ^9-THC aldehyde, **26**) was isolated [34]. These 10 Δ^9-THC-type cannabinoids are shown in Figure 3.4.

Eight cannabinoid esters, namely α-fenchyl-Δ^9-tetrahydrocannabinolate (**27**), β-fenchyl-Δ^9-tetrahydrocannabinolate (**28**), bornyl-Δ^9-tetrahydrocannabinolate (**29**), epi-bornyl-Δ^9-tetrahydrocannabinolate (**30**), α-terpenyl-Δ^9-tetrahydrocannabinolate (**31**), 4-terpenyl-Δ^9-tetrahydrocannabinolate (**32**), α-cadinyl-Δ^9-tetrahydrocannabinolate (**33**), and γ-eudesmyl-Δ^9-tetrahydrocannabinolate (**34**) were isolated from a hexane extract of a high potency variety of *C. sativa* using a variety of chromatographic techniques, including VLC, C_{18} semipreparative HPLC, and chiral HPLC [17]. These esters are shown in Figure 3.5.

Three cannabinoid-type compounds were isolated and identified in 2015, namely 8α-hydroxy-(–)-Δ^9-*trans*-THC (**35**), 8β-hydroxy-(–)-Δ^9-*trans*-THC (**36**), and 11-acetoxy-(–)-Δ^9-*trans*-THC (**37**) from high potency *C. sativa* using multiple chromatographic techniques, including silica gel VLC, C_{18}-solid-phase extraction, and HPLC [35]. In addition, the same group also isolated 8-oxo-(–)-Δ^9-*trans*-THC (**38**) from the same cannabis variety [30]. Through flash silica gel column chromatography, cannabisol (**39**) was isolated from a group of illicit cannabis samples received under the cannabis potency monitoring program at the University of Mississippi. The GC–MS analysis of these samples indicated the dimeric nature of the compounds due to the presence of two molecular peaks at *m/z* 314 and *m/z* 328, corresponding to Δ^9-THC (**17**) and 2-methyl-Δ^9-THC, respectively [36]. The structures for these cannabinoids can be found in Figure 3.6.

Two new Δ^9-THC-type cannabinoids were isolated from a medicinal cannabis variety (Italian FM2), namely (–)-Δ^9-*trans*-tetrahydrocannabiphorol (Δ^9-THCP, **40**) and (–)-Δ^9-*trans*-tetrahydrocannabihexol (Δ^9-THCH, **41**). These cannabinoids were shown to induce a variety of effects in cannabinoid tetrad pharmacological test, including

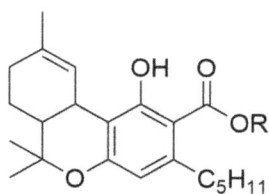

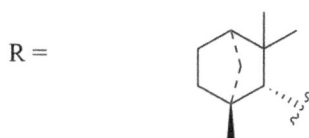

27 α-fenchyl-Δ⁹-tetrahydrocannabinolate

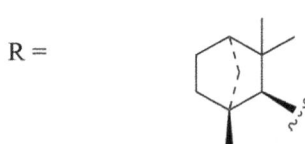

28 β-fenchyl-Δ⁹-tetrahydrocannabinolate

29 bornyl-Δ⁹-tetrahydrocannabinolate

30 epi-bornyl-Δ⁹-tetrahydrocannabinolate

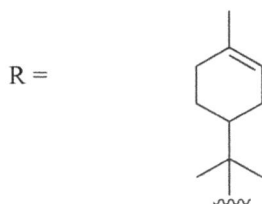

31 α-terpenyl-Δ⁹-tetrahydrocannabinolate

32 4-terpenyl-Δ⁹-tetrahydrocannabinolate

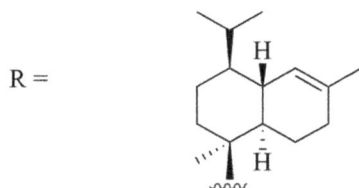

33 α-cadinyl-Δ⁹-tetrahydrocannabinolate

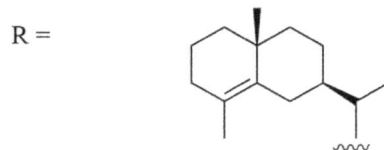

34 γ-eudesmyl-Δ⁹-tetrahydrocannabinolate

Figure 3.5: Structures of eight Δ⁹-THC esters (**27–34**).

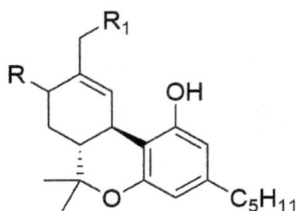

35	8α-OH-Δ^9-THC,	R=α–OH,	R_1=H
36	8β-OH-Δ^9-THC,	R=β-OH,	R_1=H
37	11-acetoxy-Δ^9-THC,	R=H,	R_1=OAc
38	8-Oxo-Δ^9-THC,	R=O,	R_1=H

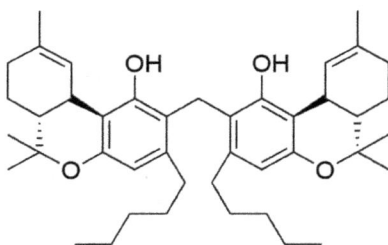

39 cannabisol

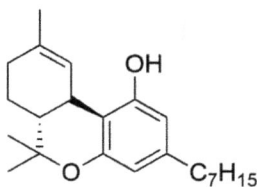

40 Δ^9-THCP

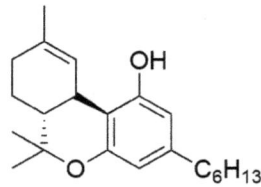

41 Δ^9-THCH

Figure 3.6: Structures of seven Δ^9-THC-type cannabinoids (**35–41**).

analgesia, catalepsy, hypomotility, and decreased rectal temperature. Multiple spectroscopic techniques were used to elucidate their chemical structures, namely ^{1}H and ^{13}C-NMR, circular dichroism (CD), and UV absorption spectroscopy, and LC–HRMS; the structure was confirmed through stereoselective synthesis. It is theorized that Δ^9-THCP (**40**) could be partially responsible for some of the pharmacological properties of multiple cannabis varieties which are problematic to assign solely to Δ^9-THC (**17**) [37, 38].

As Δ^9-THC is the active constituent of cannabis, the pharmacological studies published regarding this class of cannabinoids are numerous, numbering in the thousands. There are a few good sources, however, to have a good understanding of the pharmacology of Δ^9-THC [39–43]. On the other hand, Δ^9-THCA (A or B) was identified to have an agonistic effect on the contraction of isolated rat ileum induced by three neurotransmitters, namely acetylcholine, histamine, or barium chloride [44]. Two studies also reported the biological activity of Δ^9-THCV [31, 45] one of which determined that the effect of Δ^9-THCV in humans was about one-fourth that of Δ^9-THC.

3.4.3 Cannabitriol-type cannabinoids

There is a total of nine cannabinoids in this cannabinoid class. The structures for these cannabinoids can be seen in Figure 3.7.

The identification of cannabitriol (CBT, **42**) was reported in 1966 [46], but its chemical structure was elucidated 10 years later [47] in 1976, and its stereochemistry determined by X-ray analysis in 1984 [48]. It was reported that the specific rotation for (–)-*trans*-CBT was –107°; however, upon analyzing that of (+)-*trans*-CBT (**43**), its specific rotation turned out to be 7°, proving that the isolated (+)-*trans*-CBT (**43**) was a partially racemized mixture [48]. In 1995, all nine CBT-type cannabinoids were reported, including (±)-*cis*-CBT (**44**), (–)-*trans*-OEt-CBT (**45**), (±)-*trans*-CBT (**46**), CBT-C$_3$ (**47**), (–)-*trans*-OEt-CBT-C$_3$ (**48**), 8-OH-CBT-C$_5$ (**49**), and CBDA-C$_5$-9-O-CBT-C$_5$ ester (**50**) [49]. ElSohly et al. [50] isolated the compounds (+)-*trans*-CBT (**43**) and (–)-*trans*-ethoxy-CBT (**45**) from the ethanolic extract of cannabis, by silica gel column chromatography and identified by GC–MS. The CBT-type cannabinoid CBDA-9-O-CBT ester (**50**) was isolated in 1968 from an extract of hashish [51]. The pharmacological properties of these hydroxylated cannabinoids have not been established.

3.4.4 Δ^8-THC-type cannabinoids

There is a total of five cannabinoids in this cannabinoid class. The structures for these cannabinoids can be seen in Figure 3.8.

The cannabinoid (–)-Δ^8-*trans*-THC (**51**) was isolated in 1966 from the flowers and leaves of Maryland grown cannabis [52]. Silicic acid chromatography was used to purify Δ^8-THC from petroleum ether extract. The carboxylic acid of Δ^8-THC, namely

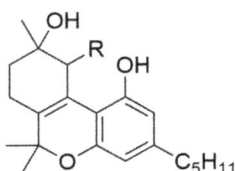

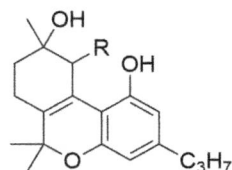

42 (-)-*trans*-CBT	R=OH
43 (+)-*trans*-CBT	R=OH
44 (±)-*cis*-CBT	R=OH
45 (-)-*trans*-CBT-OEt	R=OCH$_2$CH$_3$
46 (±)-*trans*-CBT	R=OH

| 47 CBT-C$_3$ homologue, | R=OH |
| 48 (-)-*trans*-CBT-OEt-C$_3$, | R=OCH$_2$CH$_3$ |

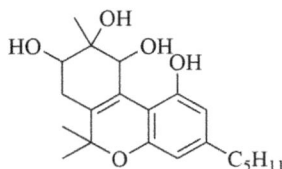

49 8-OH-CBT-C$_5$

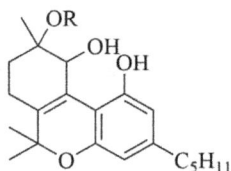

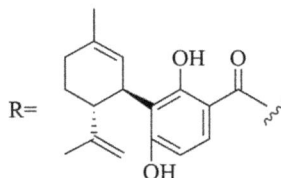

50 CBDA-C$_5$ 9-O-CBT-C$_5$ ester

Figure 3.7: Structures of nine CBT-type cannabinoids (**42–50**).

Δ^8-THCAA (**52**) was isolated from Czechoslovakian cannabis as the methyl ester [53]. Three hydroxylated Δ^8-THC-type cannabinoids were isolated from high potency *C. sativa*, namely 10α-OH-Δ^8-THC (**53**), 10β-OH-Δ^8-THC (**54**), and 10aα-OH-10-oxo-Δ^8-THC (**55**). The structure elucidation for these three cannabinoids was carried out using 1D and 2D-NMR spectral analysis [34, 35].

The pharmacological effect of Δ^8-THC (**51**) is very similar to that of Δ^9-THC (**17**), albeit less activity. Hence, there are many sources that are available describing the pharmacology associated with Δ^8-THC (**51**) [54, 55].

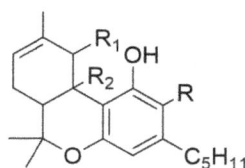

51 Δ^8-THC,	R=H,	R₁=H,	R₂=H
52 Δ^8-THCAA,	R=COOH,	R₁=H,	R₂=H
53 10α-OH-Δ^8-THC,	R=H,	R₁=α-OH,	R₂=H
54 10β-OH-Δ^8-THC,	R=H,	R₁=β-OH,	R₂=H
55 10aα-10-oxo-OH-Δ^8-THC,	R=H,	R₁=O,	R₂=OH

Figure 3.8: Structures of five Δ^8-THC-type cannabinoids (**51–55**).

3.4.5 CBN-type cannabinoids

A total of 11 cannabinoids are reported into this cannabinoid class. The structures for these compounds are shown in Figures 3.9 and 3.10.

This class of cannabinoids contains the fully aromatized derivatives of Δ^9-THC. Seven CBN-type cannabinoids, namely CBN (**56**), cannabinolic acid (CBNA, **57**), CBN-C₄ (**58**), CBN-C₃ (**59**), CBN-C₂ (**60**), cannabiorcol (CBN-C₁, **61**), and cannabinol methyl ether (CBNM, **62**) were all identified from different extracts through various chromatographic and spectral analyses [55–57]. Both 8-hydroxycannabinol (8-OH-CBN, **63**) and 8-hydroxy-cannabinolic acid A (8-OH-CBNAA, **64**) were isolated from a high potency variety of *C. sativa* in 2009 and chemically identified based on NMR and high-resolution mass spectrometry (HR–MS) analysis [20]. In addition, the isolation of 1′S-hydroxycannabinol (1′S-OH-CBN, **65**) and 4-terpenyl cannabinolate (**66**) were reported from the same high potency *C. sativa* variety six years later; their chemical structures were elucidated via GC–MS analysis [35].

The pharmacology associated with CBN (**56**) has been studied extensively over the years. It has shown to have a multitude of properties, including immunological properties [58], anticonvulsant activity [59], and anti-inflammatory activity [60]. CBN (**56**) has also shown to potentiate the effects of Δ^9-THC (**17**) [61, 62]; on the other hand, its activity is approximately only 10% of that of Δ^9-THC (**17**). There are also many other studies that deal with the pharmacological activity of CBN (**56**) [40, 54, 55, 63].

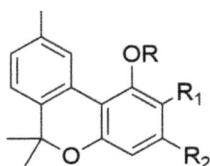

56 CBN,	R=H,	R_1=H,	R_2=C_5H_{11}
57 CBNA,	R=H,	R_1=COOH,	R_2=C_5H_{11}
58 CBN-C_4,	R=H,	R_1=H,	R_2=C_4H_9
59 CBN-C_3,	R=H,	R_1=H,	R_2=C_3H_7
60 CBN-C_2,	R=H,	R_1=H,	R_2=C_2H_5
61 CBN-C_1,	R=H,	R_1=H,	R_2=CH_3
62 CBNM,	R=CH_3,	R_1=H,	R_2=C_5H_{11}

Figure 3.9: Structures of seven CBN-type cannabinoids (**56–62**).

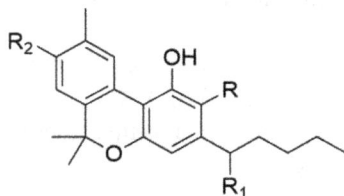

63 8-OH-CBN,	R=H,	R_1=H,	R_2=OH
64 8-OH-CBNA,	R=COOH,	R_1=H,	R_2=OH
65 1'S-OH-CBN,	R=H,	R_1=α-OH,	R_2=H

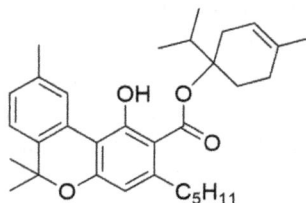

66 4-terpenyl-cannabinolate

Figure 3.10: Structures of four CBN-type cannabinoids (**63–66**).

3.4.6 CBD-type cannabinoids

A total of 13 cannabinoids are identified in this class, the structures of which are shown in Figure 3.11.

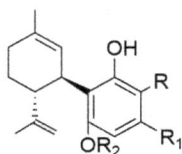

67 CBD,	R=H,	R_1=C$_5$H$_{11}$,	R_2=H
68 CBDA,	R=COOH,	R_1=C$_5$H$_{11}$,	R_2=H
69 CBDV,	R=H,	R_1=C$_3$H$_7$,	R_2=H
70 CBDM,	R=H,	R_1=C$_5$H$_{11}$,	R_2=CH$_3$
71 CBD-C$_1$,	R=H,	R_1=CH$_3$,	R_2=H
72 CBD-C$_4$,	R=H,	R_1=C$_4$H$_9$,	R_2=H
73 CBDVA,	R=COOH,	R_1=C$_3$H$_7$,	R_2=H
74 CBDP,	R=H,	R_1=C$_7$H$_{15}$,	R_2=H
75 CBDH,	R=H,	R_1=C$_6$H$_{13}$,	R_2=H

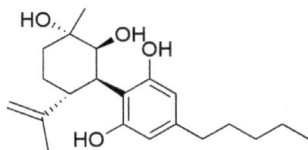

76   1,2-dihydroxycannabidiol

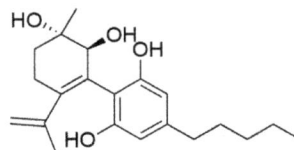

77   3,4-dehydro-1,2-dihydroxycannabidiol

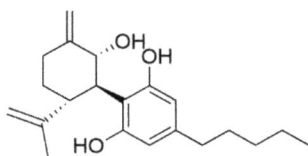

78 Hexocannabitriol

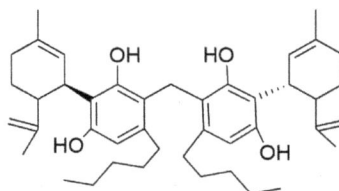

79 CBDD

Figure 3.11: Structures of 10 CBD-type cannabinoids (**67–79**).

The two major cannabinoids of the nonpsychotropic (fiber-type) variety of *C. sativa* are namely CBD (**67**) and CBDA (**68**). CBD (**67**) was first isolated in 1940; after allowing the oily CBD (**67**) to sit for several weeks, it was then crystallized [64]. CBDA (**68**) was isolated in 1955 [65]. In 1969, the complete synthesis of CBD (**67**) was reported, along with the designation of the absolute configuration as (–)-*trans*-(1 *R*,6 *R*) [66]. In the same year, cannabidivarin (CBDV, **69**), the C_3 analogue of CBD (**67**), was isolated from the ligroin extract of hashish via silica gel column chromatography [67]. Cannabidiol monomethyl ether (**70**) was isolated in 1972 from the ethanolic extract of the leaves from the Minamioshihara No. 1 (M-1) variety of *C. sativa* [68]. Also, in 1972, the C_1 analogue of CBD (**67**), namely cannabidiorcol (CBD-C_1, **71**), was isolated from a hexane extract of Lebanese hashish [27]. In 1976, cannabidiol-C_4 (CBD-C_4, **72**) was identified via its mass and lack of methylene unit [33]. Cannabidivarinic acid (CBDVA, **73**) was isolated from a benzene extract of Thailand cannabis in 1977 [16].

Two new CBD-type cannabinoids were isolated from a medicinal cannabis variety (Italian FM2), namely cannabidiphorol (CBDP, **74**) and cannabidihexol (CBDH, **75**). CBDP (**74**) has a C_7 alkyl side chain, while CBDH (**75**) has a C_6 alkyl side chain, two and one carbon units longer than the C_5 chain present in CBD (**67**), respectively. They were purified using a mixture of acetonitrile:0.1% aqueous formic acid 70:30 (v/v) as the mobile phase using a semipreparative C_{18} HPLC column . In addition, their structures were determined and elucidated using a combination of different spectroscopic techniques (^{1}H-NMR, ^{13}C-NMR, UV, and HRESIMS) and confirmed via stereoselective synthesis [37, 38].

From a hemp extract, using chromatographic purifications, three new compounds were isolated in the pure form. Their structures were elucidated as 1,2-dihydroxycannabidiol (**76**), as a pale yellow amorphous solid, 3,4-dehydro-1,2-dihydroxycannabidiol (**77**), the 3,4-unsaturated analogue of **76**, was also isolated in small amounts as a pale yellow amorphous solid, and the derivative 2α-hydroxy-$\Delta^{1,7}$-hexahydrocannabinodiol (Hexocannabitriol, **78**), is proposed, was isolated as an optically active pale yellow amorphous solid [69]. The structures of which are shown in Figure 3.11.

Last year, a dimer of CBD, namely cannabitwinol (CBDD, **79**) was isolated from the hexane extract of hemp, which was chromatographed using a silica-gel column and eluted with hexane/DCM and semipreparative C_{18}-HPLC. The structure was determined using 1D and 2D-NMR and confirmed via HREIMS and tandem mass spectrometry. The compound was identified to be a dimer of CBD through a methylene bridge; the dimerization of CBD, could be resulting from an enzymatic reaction involving a one carbon donor enzyme like methylene tetrahydrofolate [70]. The structure of CBDD (**79**) can be seen in Figure 3.11.

Rat brain cortex slices were used to show the inhibiting effect of CBD (**67**) on the incorporation of both radio carbon-labeled leucine and uridine into the proteins and nucleic acids of the slices, respectively [26]. In addition, CBD (**67**) has also shown considerable anti-inflammatory [71] and anticonvulsant [41, 72–74] activity in a multitude of studies. CBDA (**68**), on the other hand, was shown to have antibacterial activity on several microorganisms [75] as well as against *Staphylococcus aureus* [76].

3.4.7 Cannabinodiol (CBND)-type cannabinoids

There are only two cannabinoids in this cannabinoid class, the structures of which are shown in Figure 3.12.

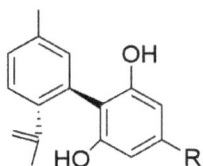

80 **CBND-C$_3$**,	R=C$_3$H$_7$
81 **CBND-C$_5$**,	R=C$_5$H$_{11}$

Figure 3.12: Structures of two CBND-type cannabinoids (**80–81**).

In 1972, cannabinodivarin (CBND-C$_3$, **80**) was detected in hashish by GC–MS analysis; five years later, cannabinodiol (CBND, **81**) was isolated from Lebanese hashish through silica gel column chromatography [77]; five years later, the structural elucidation of CBND (**81**) was determined through ^{1}H-NMR and confirmed via the photochemical transformation of CBN [78].

3.4.8 CBE-type cannabinoids

There are five cannabinoids identified in this class, the structures of which are shown in Figure 3.13.

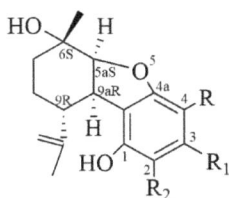

82 **CBE**,	R=H,	R$_1$=C$_5$H$_{11}$,	R$_2$=H
83 **CBEAA**,	R=COOH,	R$_1$=C$_5$H$_{11}$,	R$_2$=H
84 **CBEAB**,	R=H,	R$_1$=C$_5$H$_{11}$,	R$_2$=COOH
85 **CBE-C$_3$**,	R=H,	R$_1$=C$_3$H$_7$,	R$_2$=H
86 **CBEAB-C$_3$**,	R=H,	R$_1$=C$_3$H$_7$,	R$_2$=COOH

Figure 3.13: Structures of five CBE-type cannabinoids (**82–86**).

The parent cannabinoid of this class, namely CBE (**82**), was isolated in 1973 [79]; the configuration of the cannabinoid was determined to be 5*aS*, 6*S*, 9 *R*, and 9a*R* one year later [80]. There is a total of two acids associated with CBE (**82**), namely cannabielsoic acid A (CBEAA, **83**) and cannabielsoic acid B (CBEAB, **84**). The structure elucidation for CBEAA (**83**) and CBEAB (**84**) was carried out using both NMR spectroscopy and chemical transformations [81]. The C_3 analogue of CBE, CBE-C_3 (**85**), as well as the acidic C_3 analogue of CBEB, cannabielsoic acid B-C_3 (CBEAB-C_3, **86**), were both identified from cannabis in 1978 [82].

3.4.9 CBC-type cannabinoids

There are nine cannabinoids identified in this class, the structures of which are shown in Figure 3.14.

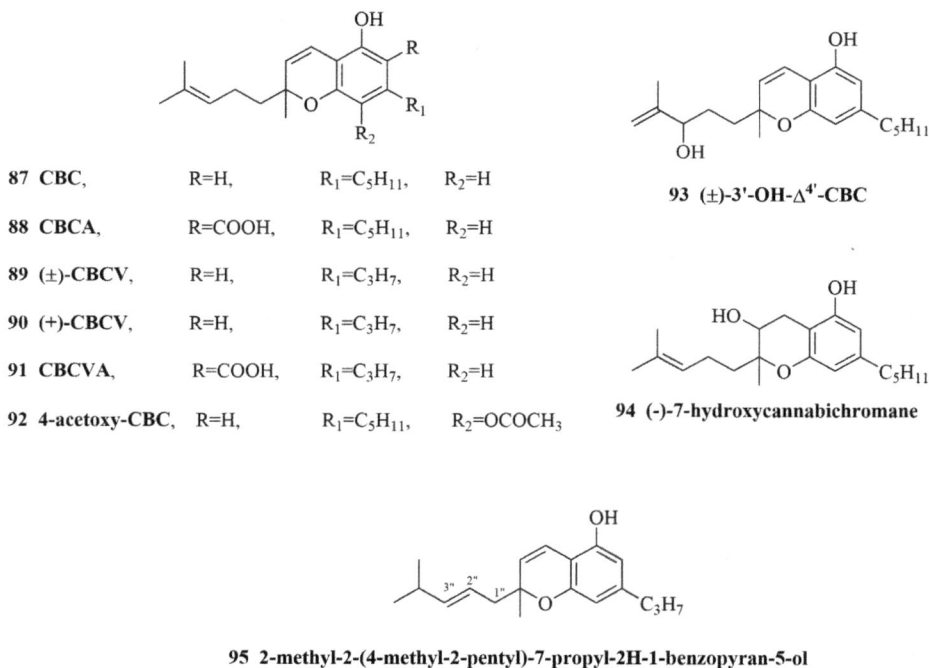

87	CBC,	R=H,	$R_1=C_5H_{11}$,	$R_2=H$
88	CBCA,	R=COOH,	$R_1=C_5H_{11}$,	$R_2=H$
89	(±)-CBCV,	R=H,	$R_1=C_3H_7$,	$R_2=H$
90	(+)-CBCV,	R=H,	$R_1=C_3H_7$,	$R_2=H$
91	CBCVA,	R=COOH,	$R_1=C_3H_7$,	$R_2=H$
92	4-acetoxy-CBC,	R=H,	$R_1=C_5H_{11}$,	$R_2=OCOCH_3$

93 (±)-3'-OH-$\Delta^{4'}$-CBC

94 (-)-7-hydroxycannabichromane

95 2-methyl-2-(4-methyl-2-pentyl)-7-propyl-2H-1-benzopyran-5-ol

Figure 3.14: Structures of nine CBC-type cannabinoids (**87–95**).

One of the first cannabinoids to be isolated in this class, cannabichromene (CBC, **87**), was isolated from a hexane extract of hashish in 1966 through Florisil column chromatography [83]. The carboxylic acid of CBC, namely cannabichromenic acid (CBCA, **88**), was isolated from a benzene extract of hemp two years later [84]. Just two years after this, the racemic mix of the C_3 homologue of CBC, Cannabichromevarin ((±)-CBCV, **89**) was identified through GC–MS analysis. In the same report, the (R) isomer of CBCV, namely ((+)-CBCV, **90**) and cannabichromevarinic acid (CBCVA, **91**), were also isolated from "Meao" cannabis variety from Thailand [16]. In 1984, a CBC-C_3 derivative was reported, namely 2-methyl-2-(4-methyl-2-pentyl)-7-propyl-2 H-1-benopyran-5-ol (**92**); the location of the double bond in the isoprenyl side chain shifted positions from a C_3–C_4 double bond to a C_2–C_3 double bond [85]. Three other CBC derivatives, 4-acetoxy-CBC (**93**), (±)-3'-hydroxy-$\Delta^{4'}$-CBC (**94**), and (–)-7-hydroxycannabichromane (**95**), were isolated from high potency *C. sativa* using a variety of chromatographic methodologies, including silica gel VLC, NP-HPLC, and C_{18}-HPLC. The spectral identification of these compounds was carried out using 1D and 2D-NMR spectroscopic techniques [20].

CBC (**87**) was shown to have a sedative and ataxia effect in dogs [86]; however, it showed no activity in human smoking experiments [87] or in the Rhesus monkey [88]. In rats, CBC (**87**) was shown to induce a slight loss of muscular coordination as well as an increase in urination and cyanosis [89]. The interaction between CBC (**87**) and Δ^9-THC (**17**) has been theorized in a study using monkeys in 1978 [90].

3.4.10 CBL-type cannabinoids

There are three cannabinoids in this class, the structures of which are shown in Figure 3.15.

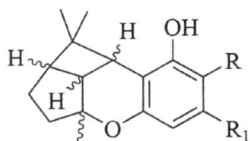

96	**CBL**,	R=H,	R_1=C_5H_{11}
97	**CBLA**,	R=COOH,	R_1=C_5H_{11}
98	**CBLV**,	R=H,	R_1=C_3H_7

Figure 3.15: Structures of three CBL-type cannabinoids (**96–98**).

Cannabicyclol (CBL, **96**) was isolated from hashish in 1967 [91]. The relative configuration of this cannabinoid was determined three years later by X-ray analysis (**96**) [92]. In 1972, the carboxylic acid of CBL, namely CBLA (**97**) and its C_3 homologue, cannabicylovarin (CBLV, **98**) were both isolated as optically inactive crystals [27]. CBLA (**97**) was determined to be produced during the natural irradiation of CBCA (**88**); this demonstrates that CBLA (**97**) is not a natural cannabinoid [68].

3.4.11 Miscellaneous-type cannabinoids

There are 31 cannabinoids which are reported in this cannabinoid class. The structures for these compounds are shown in Figures 3.16–3.18.

In 1971, cannabicitran (**99**) was synthesized and named citrylidene-cannabis [93], however, this cannabinoid was first isolated from Lebanese hashish three years later, in 1974 [94]. Four miscellaneous-type cannabinoids were isolated in 1975, namely dehydrocannabifuran (**100**), cannabifuran (**101**), cannabichromanone (CBCN-C$_5$, **102**), and 10-oxo-$\Delta^{6a(10a)}$-tetrahydrocannabinol (OTHC, **103**) [95]. The *cis*-isomer of Δ^9-THC, namely (–)-*cis*-Δ^9-THC (**104**), was identified in confiscated marijuana samples in 1977 [96]. CBCN-C$_3$ (**105**) and cannabicoumaranone-C$_5$ (CBCON, **106**) were both isolated in 1978; however, their absolute configurations are not known [82, 97]. Then, the following year, (–)-cannabiripsol ((–)-CBR, **107**) was isolated from a cannabis variety from South America [98]. Cannabiglendol-C$_3$ (8-OH-*iso*-HHCV-C$_3$, **108**) was isolated in 1981 from an Indian cannabis variety; the cannabinoid was identified through spectral means as well as by comparing it to the synthetic C$_5$ homologue prepared [99]. The same year, another miscellaneous-type cannabinoid, namely (±)-Δ^7-*cis-iso*-THCV-C$_3$ (**109**) was isolated [100]. In 1984, three new miscellaneous-type compounds were isolated, namely (–)-cannabitetrol (**110**) [93], (–)-Δ^7-*trans*-(1 *R*,3 *R*,6 *R*)-*iso*-THCV-C$_3$ (**111**), and (–)-Δ^7-*trans*-(1 *R*,3 *R*,6 *R*)-*iso*-tetrahydrocannabinol-C$_5$ (**112**) [101]. The structures of these miscellaneous-type cannabinoids can be found in Figure 3.16.

In 2008, six new miscellaneous-type cannabinoids were isolated from a high-potency variety *C. sativa*. Three of these cannabinoids were variants of CBCNs, namely CBCN B (**113**), CBCN C (**114**), and CBCN D (**115**) [102]. The absolute configuration of these three cannabinoids was assigned based on inspection of their CD spectra as well as on the basis of the Mosher ester. These cannabinoids were isolated using semipreparative RP-HPLC. The other three cannabinoids, namely (–)-(7 *R*)-cannabicoumaronic acid (**116**), 4-acetoxy-2-geranyl-5-hydroxy-3 *n*-pentylphenol (**117**), and 2-geranyl-5-hydroxy-3 *n*-petnyl-1,4-benzoquinone (**118**), were isolated from the buds and leaves of the cannabis plant. These three cannabinoids were isolated through a variety of chromatographic techniques, including silica-gel VLC, solid-phase extraction, RP columns, and NP-HPLC [21]. In the same year, the same group also isolated another miscellaneous-type cannabinoid namely 5-acetoxy-6-geranyl-3-*n*-pentyl-1,4-benzoquinone (**119**) [19]. Two years later, a new cannabinoid named cannabimovone (**120**) was isolated from a nonpsychotropic *C. sativa* variety. This cannabinoid is presumably a CBD metabolite degradation product and was isolated from a polar fraction of hemp using flash chromatography over RP silica-gel followed by NP-HPLC [103]. In 2022, for the first time anhydrocannabimovone (**121**), isolated for the first time as a natural product from mother liquors obtained from crystallization of CBD from hemp [69].

A tetracyclic cannabinoid, cannabioxepane (**122**) was isolated a year later from the cannabis variety called carmagnole using many chromatographic techniques, including RP silica-gel column, NP-silica gel column, and NP-HPLC chromatography [104]. The structures of compound **113–122** are shown in Figure 3.17.

In 2015, the same group reported the isolation of seven new miscellaneous-type cannabinoids, the structures of which are shown in Figure 3.18. These compounds were structurally determined to be 10α-hydroxy-$\Delta^{9,(11)}$-hexahydrocannabinol (**123**),

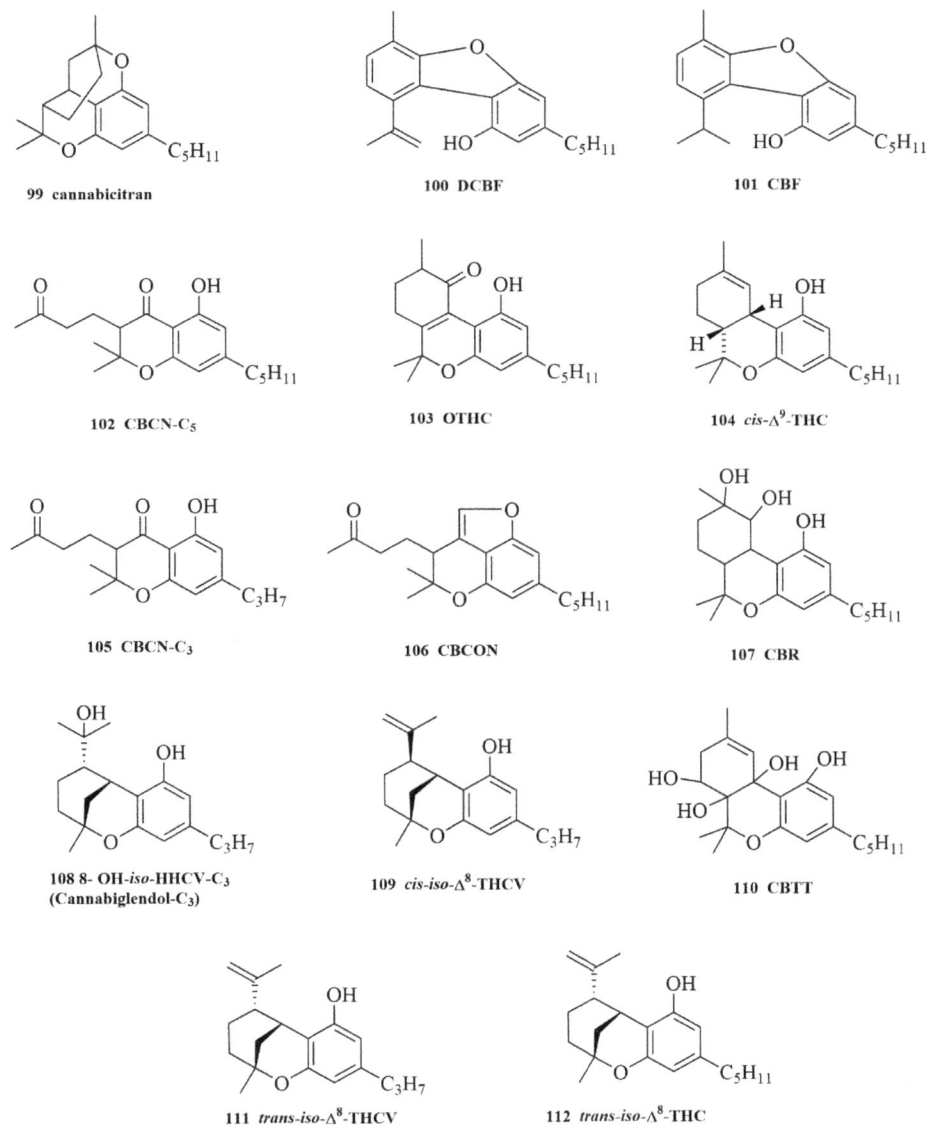

99 cannabicitran

100 DCBF

101 CBF

102 CBCN-C$_5$

103 OTHC

104 cis-Δ^9-THC

105 CBCN-C$_3$

106 CBCON

107 CBR

108 8- OH-iso-HHCV-C$_3$ (Cannabiglendol-C$_3$)

109 cis-iso-Δ^8-THCV

110 CBTT

111 trans-iso-Δ^8-THCV

112 trans-iso-Δ^8-THC

Figure 3.16: Structures of fourteen miscellaneous-type cannabinoids (**99–112**).

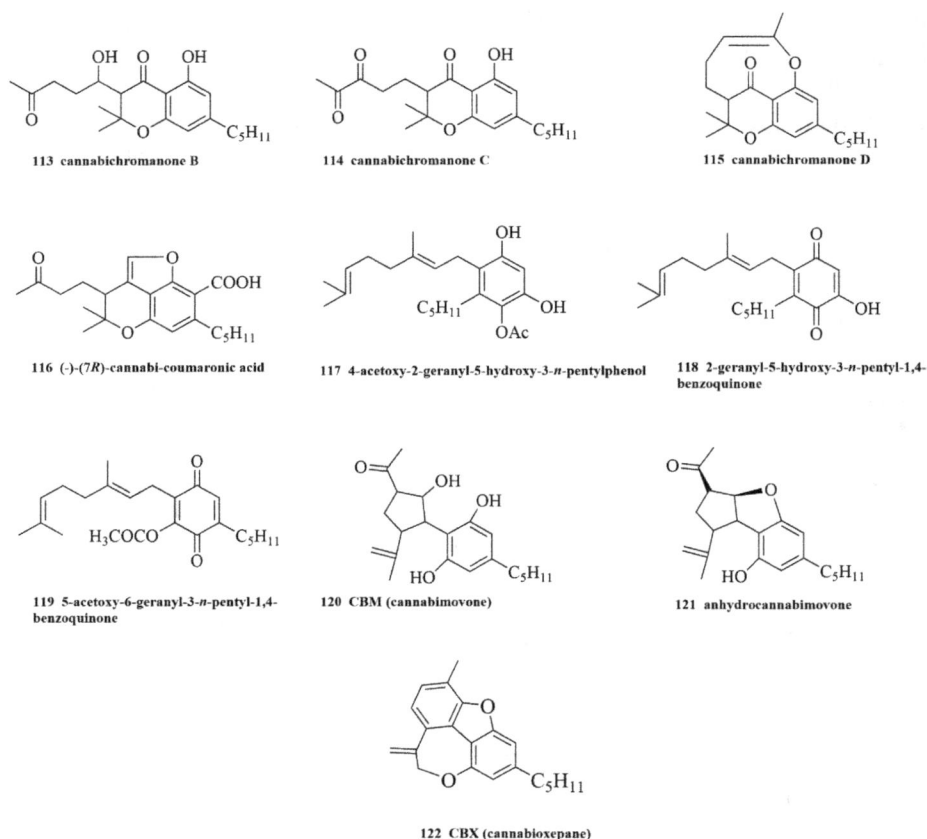

113 cannabichromanone B

114 cannabichromanone C

115 cannabichromanone D

116 (-)-(7R)-cannabi-coumaronic acid

117 4-acetoxy-2-geranyl-5-hydroxy-3-*n*-pentylphenol

118 2-geranyl-5-hydroxy-3-*n*-pentyl-1,4-benzoquinone

119 5-acetoxy-6-geranyl-3-*n*-pentyl-1,4-benzoquinone

120 CBM (cannabimovone)

121 anhydrocannabimovone

122 CBX (cannabioxepane)

Figure 3.17: Structures of nine miscellaneous-type cannabinoids (**113–122**).

9β,10β-epoxy-hexahydrocannabinol (**124**), 9α-hydroxy-hexahydrocannabinol (**125**), 7-oxo-9α-hydroxy-hexahydrocannabinol (**126**), 10α-hydroxyhexahydrocannabinol (**127**), 10α-hydroxy-hexahydrocannabinol (**128**), and 9α-hydroxy-10-oxo-$\Delta^{6a,(10a)}$-THC (**129**) [34, 35]. The structures of these cannabinoids are shown in Figure 3.18.

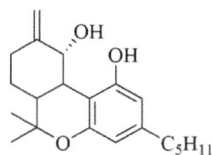

123 10α-OH-D$^{9,(11)}$-hexahydrocannabinol

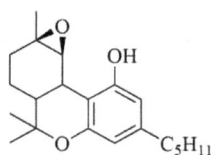

124 $9\beta,10\beta$-epoxyhydrocannabinol

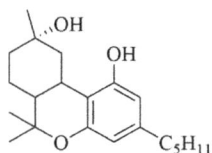

125 9α-hydroxy-hexa-hydrocannabinol

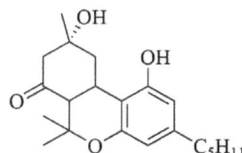

126 7-oxo-9α-hydroxy-hexahydrocannabinol

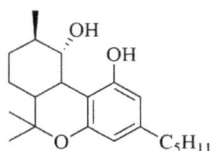

127 10α-hydroxy-hexahydrocannabinol

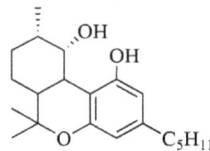

128 10α-hydroxy-hexahydrocannabinol

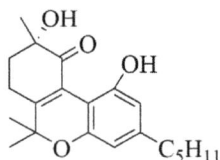

129 9α-hydroxy-10-oxo-$\Delta^{6a,(10a)}$-THC

Figure 3.18: Structures of seven miscellaneous-type cannabinoids (**123–129**).

References

[1]   Fleming MP, Clarke RC. Physical evidence for the antiquity of *Cannabis sativa* L. Journal of the International Hemp Association 1998; 5: 80–95.

[2]   Substance Abuse and Mental Health Services Administration. Key substance use and mental health indicators in the United States: Results from the 2019 National Survey on Drug Use and Health, Rockville, MD, 2020.

[3]   Bocsa I, Karus M. The cultivation of hemp: Botany, varieties, cultivation and harvesting. Hemptech: Sabastopol, CA, USA. 1998.

[4]   Li H-L. An archaeological and historical account of cannabis in China. Economic Botany 1974; 28: 437–448.

[5]   Gaoni Y, Mechoulam R. Isolation, structure, and partial synthesis of an active constituent of hashish. Journal of the American Chemical Society 1964; 86: 1646–1647.

[6]   Mechoulam R, Gaoni Y. The absolute configuration of δ1-tetrahydrocannabinol, the major active constituent of hashish. Tetrahedron Letters 1967; 8: 1109–1111.

[7]   Mechoulam R, Burstein SH. Marijuana: Chemistry, pharmacology, metabolism and clinical effects Contributors-SH Burstein [And Others], Academic Press, New York, 1973.

[8]   Radwan MM, Chandra S, Gul S, ElSohly MA. Cannabinoids, phenolics, terpenes and alkaloids of cannabis. Molecules 2021; 26: 2774.

[9]   Martin B, Mechoulam R, Razdan R. Discovery and characterization of endogenous cannabinoids. Life Sciences 1999; 65: 573–595.

[10]   Cahn R. 326. *Cannabis indica* resin. Part IV. The synthesis of some 2: 2-dimethyldibenzopyrans, and confirmation of the structure of cannabinol. Journal of the Chemical Society (Resumed) 1933; (0): 1400–1405.

[11]   Gaoni Y, Mechoulam R. Structure+ synthesis of cannabigerol new hashish constituent. In: Royal Soc Chemistry Thomas Graham House, Science Park, Milton Rd, Cambridge . . ., 1964; 82-&.

[12]   Yamauchi T, Shoyama Y, Matsuo Y, Nishioka I. Cannabigerol monomethyl ether, a new component of hemp. Chemical and Pharmaceutical Bulletin 1968; 16: 1164–1165.

[13]   Mechoulam R, Gaoni Y. Hashish – IV: The isolation and structure of cannabinolic cannabidiolic and cannabigerolic acids. Tetrahedron 1965; 21: 1223–1229.

[14]   Shoyama Y, Yamauchi T, Nishioka I. Cannabis. V. Cannabigerolic acid monomethyl ether and cannabinolic acid. Chemical and Pharmaceutical Bulletin 1970; 18: 1327–1332.

[15]   Shoyama Y, Hirano H, Oda M, Somehara T, Nishioka I. Cannabichromevarin and cannabigerovarin, two new propyl homologues of cannabichromene and cannabigerol. Chemical and Pharmaceutical Bulletin 1975; 23: 1894–1895.

[16]   Shoyama Y, Hirano H, Makino H, Umekita N, Nishioka I. Cannabis. X. The isolation and structures of four new propyl cannabinoid acids, tetrahydrocannabivarinic acid, cannabidivarinic acid, cannabichromevarinic acid and cannabigerovarinic acid, from Thai Cannabis, 'Meao variant'. Chemical and Pharmaceutical Bulletin 1977; 25: 2306–2311.

[17]   Taura F, Morimoto S, Shoyama Y. Cannabinerolic acid, a cannabinoid from *Cannabis sativa*. Phytochemistry 1995; 39: 457–458.

[18]   Pollastro F, Taglialatela-Scafati O, Allara M, Munoz E, Di Marzo V, De Petrocellis L, Appendino G. Bioactive prenylogous cannabinoid from fiber hemp (*Cannabis sativa*). Journal of Natural Products 2011; 74: 2019–2022.

[19]   Radwan MM, Ross SA, Slade D, Ahmed SA, Zulfiqar F, ElSohly MA. Isolation and characterization of new cannabis constituents from a high potency variety. Planta Medica 2008; 74: 267–272.

[20]   Radwan MM, ElSohly MA, Slade D, Ahmed SA, Khan IA, Ross SA. Biologically active cannabinoids from high-potency *Cannabis sativa*. Journal of Natural Products 2009; 72: 906–911.

[21]   Ahmed SA, Ross SA, Slade D, Radwan MM, Zulfiqar F, ElSohly MA. Cannabinoid ester constituents from high-potency *Cannabis sativa*. Journal of Natural Products 2008; 71: 536–542.

[22]   Grunfeld Y, Edery H. Psychopharmacological activity of some substances extracted from *Cannabis sativa* L. (hashish). Electroencephalography and Clinical Neurophysiology 1969; 27: 219–220.

[23]   Mechoulam R, Shani A, Edery H, Grunfeld Y. Chemical basis of hashish activity. Science 1970; 169: 611–612.

[24]   Coldwell B, Bailey K, Paul J, Anderson, G. Interaction of cannabinoids with pentobarbital in rats. Toxicology and Applied Pharmacology 1974; 29: 59–69.

[25]   Hattori T, Jakubovic A, McGeer P. The effect of cannabinoids on the number of nuclear membrane-attached ribosomes in infant rat brain. Neuropharmacology 1973; 12: 995–999.

[26]   Jakubovič A, McGeer PL. Inhibition of rat brain protein and nucleic acid synthesis by cannabinoids in vitro. Canadian Journal of Biochemistry 1972; 50: 654–662.

[27] Vree T, Breimer D, Van Ginneken C, Van Rossum J. Identification of cannabicyclol with a pentyl or propyl side-chain by means of combined as chromatography – Mass spectrometry. Journal of Chromatography 1972; 74: 124–127.

[28] Korte F, Sieper H, Tira S. New results on hashish-specific constituents. Bulletin on Narcotics 1965; 17: 35–43.

[29] Mechoulam R, Ben-Zvi Z, Yagnitinsky B, Shani A. A new tetrahydrocannabinolic acid. Tetrahedron Letters 1969; 10: 2339–2341.

[30] Gill E. Propyl homologue of tetrahydrocannabinol: Its isolation from Cannabis, properties, and synthesis. Journal of the Chemical Society 1971: 579–582.

[31] Paris M, Ghirlanda, C, Chaigneau M, Giry L. Δ^1-Tetrahydrocannabivarolic acid, new constituent of *Cannabis sativa*. C R Acad Sci Ser C 1973; 276: 205–207.

[32] Turner CE, Hadley KW, Fetterman PS, Doorenbos NJ, Quimby MW, Waller C. Constituents of *Cannabis sativa* L. IV: Stability of cannabinoids in stored plant material. Journal of Pharmaceutical Sciences 1973; 62: 1601–1605.

[33] Harvey D. Characterization of the butyl homologues of Δ1-tetrahydrocannabinol, cannabinol and cannabidiol in samples of cannabis by combined gas chromatography and mass spectrometry. Journal of Pharmacy and Pharmacology 1976; 28: 280–285.

[34] Ahmed SA, Ross SA, Slade D, Radwan MM, Khan IA, ElSohly MA. Minor oxygenated cannabinoids from high potency *Cannabis sativa* L. Phytochemistry 2015; 117: 194–199.

[35] Radwan MM, ElSohly MA, El-Alfy AT, Ahmed SA, Slade D, Husni AS, Manly SP, Wilson L, Seale S, Cutler SJ. Isolation and pharmacological evaluation of minor cannabinoids from high-potency *Cannabis sativa*. Journal of Natural Products 2015; 78: 1271–1276.

[36] Zulfiqar F, Ross SA, Slade D, Ahmed SA, Radwan MM, Ali Z, Khan IA, ElSohly MA. Cannabisol, a novel Δ^9-THC dimer possessing a unique methylene bridge, isolated from *Cannabis sativa*. Tetrahedron Letters 2012; 53: 3560–3562.

[37] Citti C, Linciano P, Russo F, Luongo L, Iannotta M, Maione S, Laganà A, Capriotti AL, Forni F, Vandelli MA. A novel phytocannabinoid isolated from *Cannabis sativa* L. with an in vivo cannabimimetic activity higher than Δ9-tetrahydrocannabinol: Δ9-Tetrahydrocannabiphorol. Scientific Reports 2019; 9: 1–13.

[38] Linciano P, Citti C, Russo F, Tolomeo F, Laganà A, Capriotti AL, Luongo L, Iannotta M, Belardo C, Maione S. Identification of a new cannabidiol n-hexyl homolog in a medicinal cannabis variety with an antinociceptive activity in mice: Cannabidihexol. Scientific Reports 2020; 10: 1–11.

[39] Mechoulam R. Marijuana: Chemistry, pharmacology, metabolism, and clinical effects. New York, Academic Press, 1973.

[40] Braude MC, Szara SI. Pharmacology of marihuana. New York, Raven Press, 1976.

[41] Nahas GG. Marihuana: Chemistry, biochemistry, and cellular effects [proceedings of the Satellite Symposium on Marihuana (Matinkylä, Finland) of the 6. Internat. Congress of Pharmacology held July 26–27, 1975 in Helsinki, Finland]. New York, Springer, 1976.

[42] Cohen S. The therapeutic potential of marihuana: Proceedings of a conference on the therapeutic potential of marihuana, held at the Asilomar Conference Center in Pacific Grove, Calif., Nov. 1975. Plenum Medical Book Comp., 1976.

[43] Kettenes-van den Bosch J, Salemink C, Van Noordwijk J, Khan I. Biological activity of the tetrahydrocannabinols. Journal of Ethnopharmacology 1980; 2: 197–231.

[44] Tampier L, Linetzky R, Mardones J. Effect of cannabinols from marihuana on smooth muscle (author's transl). Archivos de Biologia Y Medicina Experimentales 1973; 9: 16–19.

[45] Hollister LE. Structure-activity relationships in man of cannabis constituents, and homologs and metabolites of Δ9-tetrahydrocannabinol. Pharmacology 1974; 11: 3–11.

[46] Obata Y, Ishikawa Y. Studies on the Constituents of hemp plant (*Cannabis sativa* L.) Part I. Volatile phenol fraction. Journal of the Agricultural Chemical Society of Japan 1960; 24: 667–669.

[47] Chan W, Magnus K, Watson H. The structure of cannabitriol. Experientia 1976; 32: 283–284.

[48] McPhail AT, ElSohly HN, Turner CE, ElSohly MA. Stereochemical assignments for the two enantiomeric pairs of 9, 10-dihydroxy-Δ6a (10a)-tetrahydrocannabinols. X-Ray crystal structure analysis of (±) *trans*-cannabitriol. Journal of Natural Products 1984; 47: 138–142.

[49] Ross S, ElSohly M. Constituents of *Cannabis sativa* L. XXVIII A review of the natural constituents: 1980–1994. Zagazig Journal of Pharmaceutical Sciences 1995; 4: 150–160.

[50] ElSohly M, El-Feraly F, Turner C. Isolation and characterization of (+) cannabitriol and (-)-10 ethoxy 9 hydroxy delta 6a tetrahydrocannabinol: Two new cannabinoids from *Cannabis sativa* L. extract. Lloydia, 1977; 40: 275–280.

[51] Claussen U, Von Spulak F, Korte F. Haschisch – XIV: Zur kenntnis der inhaltsstoffe des haschisch. Tetrahedron 1968; 24: 1021–1023.

[52] Hively RL, Mosher WA, Hoffmann FW. Isolation of *trans*-Δ6-tetrahydrocannabinol from marijuana. Journal of the American Chemical Society 1966; 88: 1832–1833.

[53] Krejcí Z, Šantavý F. Isolation of two new cannabinoid acids from *Cannabis sativa* L. of Czechoslovak origin. Acta Universitatis Palackianae Olomucensis Facultatis Medicae 1975; 74: 161–166.

[54] Smith SG, Waller CW, Johnson JJ, Buelke JUDY, Turner C. Marihuana: An annotated bibliography (Book Review). The Psychological Record 1977; 27: 365.

[55] Neumeyer JL, Shagoury RA. Chemistry and pharmacology of marijuana. Journal of Pharmaceutical Sciences 1971; 60: 1433–1457.

[56] Turner CE, Elsohly MA, Boeren EG. Constituents of *Cannabis sativa* L. XVII. A review of the natural constituents. Journal of Natural Products 1980; 43: 169–234.

[57] Bercht C, Kuppers F, Lousberg R. Volatile Constituents of *Cannabis sativa* L.; UN Secretariat Document. In: ST/SUA/SER. 5/29, 22 July 1971; 1971.

[58] Cushman P, Jr. Cannabinols and the rosette forming properties of lymphocytes in vitro. Life Sciences 1976; 19: 875–885.

[59] Karler R, Cely W, Turkanis SA. The anticonvulsant activity of cannabidiol and cannabinol. Life Sciences 1973; 13: 1527–1531.

[60] Sofia RD, Delgado CJ, Douglas JF. The effects of various naturally occurring cannabinoids on hypotonic-hyperthermic lysis of rat erythrocytes. European Journal of Pharmacology 1974; 27: 155–157.

[61] Karniol I, Carlini E. The content of (-) Δ^9-*trans*-tetrahydrocannabinol (Δ^9-THC) does not explain all biological activity of some Brazilian marihuana samples. Journal of Pharmacy and Pharmacology 1972; 24: 833–835.

[62] Musty R, Karniol I, Shirikawa I, Takahashi R, Knobel E. Interactions of delta-9-tetrahydrocannabinol and cannabinol in man. The Pharmacology of Marihuana 1976; 2: 559–563.

[63] ElSohly M, Turner C. A review of nitrogen containing compounds from *Cannabis sativa* L. Pharmaceutisch Weekblad 1976; 3: 1069–1075.

[64] Adams R, Hunt M, Clark J. Structure of cannabidiol, a product isolated from the marihuana extract of Minnesota wild hemp. I. Journal of the American Chemical Society 1940; 62: 196–200.

[65] Krejci Z, Santavy F. Isolation of other substances from the leaves of Indian hemp. Acta Univ Palacki Olomuc 1955; 6: 59.

[66] Petrzilka T, Haefliger W, Sikemeier C. Synthesis of hashish components. IV. Helvetica Chimica Acta 1969; 52: 1102–1134.

[67] Vollner L, Bieniek D, Korte F. Hashish. XX. Cannabidivarin, a new hashish constituent. Tetrahedron Letters, 1969; 3: 145–147.

[68] Shoyama Y, Oku R, Yamauchi T, Nishioka I. Cannabis. VI. Cannabicyclolic acid. Chemical and Pharmaceutical Bulletin 1972; 20: 1927–1930.

[69]   Chianese G, Sirignano C, Benetti E, Marzaroli V, Collado JA, de la Vega L, Appendino G, Muñoz E,
       Taglialatela-Scafati O. A Nrf-2 stimulatory hydroxylated cannabidiol derivative from hemp (*Cannabis
       sativa*). Journal of Natural Products 2022; 85: 1089–1097.
[70]   Chianese G, Lopatriello A, Schiano-Moriello A, Caprioglio D, Mattoteia D, Benetti E, Ciceri D, Arnoldi
       L, De Combarieu E, Vitale RM. Cannabitwinol, a dimeric phytocannabinoid from hemp, *Cannabis
       sativa* L., is a selective thermo-TRP modulator. Journal of Natural Products 2020; 83: 2727–2736.
[71]   Atalay S, Jarocka-Karpowicz I, Skrzydlewska E. Antioxidative and anti-inflammatory properties of
       cannabidiol. Antioxidants 2020; 9: 21.
[72]   Izquierdo I. Effect of anticonvulsant drugs on the number of afferent stimuli needed to cause a
       hippocampal seizure discharge. Pharmacology 1974; 11: 146–150.
[73]   Izquierdo I, Orsingher OA, Berardi AC. Effect of cannabidiol and of other *Cannabis sativa* compounds
       on hippocampal seizure discharges. Psychopharmacologia 1973; 28: 95–102.
[74]   Carlini E, Leite J, Tannhauser M, Berardi A. Cannabidiol and *Cannabis sativa* extract protect mice and
       rats against convulsive agents. Journal of Pharmacy and Pharmacology 1973; 25: 664–665.
[75]   Abichabki N, Zacharias LV, Moreira NC, Bellissimo-Rodrigues F, Moreira FL, Benzi JR, Ogasawara T,
       Ferreira JC, Ribeiro CM, Pavan FR. Potential cannabidiol (CBD) repurposing as antibacterial and
       promising therapy of CBD plus polymyxin B (PB) against PB-resistant Gram-negative bacilli.
       Scientific Reports 2022; 12: 1–15.
[76]   Van Klingeren B, Ten Ham M. Antibacterial activity of Δ9-tetrahydrocannabinol and cannabidiol.
       Antonie, Van Leeuwenhoek 1976; 42: 9–12.
[77]   Van Ginneken C, Vree T, Breimer D, Thijssen H, Van Rossum J. Cannabinodiol, a new hashish
       consituent, identified by gaschromatography-mass spectrometry. https://repository.ubn.ru.nl/bit
       stream/handle/2066/143107/143107.pdf 1973.
[78]   Ch LRJ, Bercht CL, van Ooyen R, Spronck HJ. Cannabinodiol: Conclusive identification and synthesis
       of a new cannabinoid from *Cannabis sativa*. Phytochemistry 1977; 16: 595–597.
[79]   Bercht C, Lousberg R, Küppers F, Salemink C, Vree T, Van Rossum J. Cannabis: VII. Identification of
       cannabinol methyl ether from hashish. Journal of Chromatography A 1973; 81: 163–166.
[80]   Uliss DB, Razdan RK, Dalzell HC. Stereospecific intramolecular epoxide cleavage by phenolate anion.
       Synthesis of novel and biologically active cannabinoids. Journal of the American Chemical Society
       1974; 96: 7372–7374.
[81]   Shani A, Mechoulam R. Cannabielsoic acids: Isolation and synthesis by a novel oxidative cyclization.
       Tetrahedron 1974; 30: 2437–2446.
[82]   Grote H, Spiteller G. New cannabinoids. 2. Journal of Chromatography 1978; 154: 13–23.
[83]   Gaoni Y, Mechoulam R. Cannabichromene, a new active principle in hashish. Chemical
       Communications (London), 1966; 20–21.
[84]   Shoyama Y, Fujita T, Yamauchi T, Nishioka I. Cannabichromenic acid, a genuine substance of
       cannabichromene. Chemical and Pharmaceutical Bulletin 1968; 16: 1157–1158.
[85]   Morita M, Ando H. Analysis of hashish oil by gas chromatography/mass spectrometry. Kagaku
       Keisatsu Kenkyusho Hokoku Hokagaku Hen 1984; 37: 137–140.
[86]   Razdan RK. Structure-activity relationships. NIDA Research Monograph 1976; 79: 3.
[87]   Isbell H, Gorodetzsky C, Jasinski D, Claussen U, Spulak F, Korte F. Effects of (−) Δ9-trans-
       tetrahydrocannabinol in man. Psychopharmacologia 1967; 11: 184–188.
[88]   Rosenkrantz H, Fleischman RW, Grant RJ. Toxicity of short-term administration of cannabinoids to
       rhesus monkeys. Toxicology and Applied Pharmacology 1981; 58: 118–131.
[89]   Razdan R, Pars H. Studies on Cannabis constituents and synthetic analogues. Botany and Chemistry
       of Cannibis 1970: 137–149
[90]   Garey R, Turner C, Heath R. Administration of marihuana smoke and marihuana constituents to
       monkeys: Methods, effects, and problems. Bulletin on Narcotics 1980; 32: 55–66.

[91] Mechoulam R, Gaoni Y. Recent advances in the chemistry of hashish. Fortschritte der Chemie Organischer Naturstoffe/Progress in the Chemistry of Organic Natural Products/Progrès dans la Chimie des Substances Organiques Naturelles. In Progress in the Chemistry of Organic Natural Products; Zechmeister, L., Ed.; Springer: Vienna, Austria, 1967; 25: 175–213.

[92] Begley M, Clarke D, Crombie L, Whiting D. The x-ray structure of dibromocannabicyclol: Structure of bicyclomahanimbine. Journal of the Chemical Society D: Chemical Communications 1970; 22: 1547–1548.

[93] Crombie L, Ponsford R. Synthesis of cannabinoids by pyridine-catalysed citral–olivetol condensation: Synthesis and structure of cannabicyclol, cannabichromen, (hashish extractives), citrylidene-cannabis, and related compounds. Journal of the Chemical Society C: Organic 1971; 4: 796–804.

[94] Bercht CL, Lousberg RJC, Küppers FJ, Salemink CA. Cannabicitran: A new naturally occurring tetracyclic diether from Lebanese Cannabis sativa. Phytochemistry 1974; 13: 619–621.

[95] Friedrich-Fiechtl J, Spiteller G. Neue cannabinoide – 1. Tetrahedron 1975; 31: 479–487.

[96] Smith RM, Kempfert KD. Delta1-3, 4 cis tetrahydrocannabinol in Cannabis sativa. Phytochemistry 1977; 16: 1088–1089.

[97] Grote H, Spiteller G. New cannabinoids. The structure of cannabicoumaronone and analogous compounds. Chemischer Informationsdienst 1979; 10: no–no.

[98] Boeren E, ElSohly M, Turner C. Cannabiripsol: A novel Cannabis constituent. Experientia 1979; 35: 1278–1279.

[99] Turner C, Mole M, Hanus L, Elsohly H. Constituents of Cannabis sativa. XIX. Isolation and structure elucidation of cannabiglendol, a novel cannabinoid from an Indian variant. Journal of Natural Products 1981; 44: 27–33.

[100] Shoyama Y, Morimoto S, Nishioka I. Cannabis. XIV. Two new propyl cannabinoids, cannabicyclovarin and Δ^7-cis-iso-tetrahydrocannabivarin, from Thai cannabis. Chemical and Pharmaceutical Bulletin 1981; 29: 3720–3723.

[101] ElSohly HN, Boeren EG, Turner CE, ElSohly MA. Constituents of Cannabis sativa L. XXIIII: Cannabitetrol, a new polyhydroxylated cannabinoid. Orlando, FL, Academic Press, Inc., 1984.

[102] Ahmed SA, Ross SA, Slade D, Radwan MM, Khan IA, ElSohly MA. Structure determination and absolute configuration of cannabichromanone derivatives from high potency Cannabis sativa. Tetrahedron Letters 2008; 49: 6050–6053.

[103] Taglialatela-Scafati O, Pagani A, Scala F, De Petrocellis L, Di Marzo V, Grassi G, Appendino G. Cannabimovone, a cannabinoid with a rearranged terpenoid skeleton from hemp. Wiley Online Library, 2010.

[104] Pagani A, Scala F, Chianese G, Grassi G, Appendino G, Taglialatela-Scafati O. Cannabioxepane, a novel tetracyclic cannabinoid from hemp, Cannabis sativa L. Tetrahedron 2011; 67: 3369–3373.

Waseem Gul, Shahbaz W. Gul and Mahmoud A. ElSohly*

4 The non-cannabinoid constituents of *Cannabis sativa*

Abstract: Although cannabinoids are the most studied constituents of *Cannabis sativa* L. *C. sativa*, there have been over 400 non-cannabinoid constituents which have been identified from *C. sativa*. These non-cannabinoids can be classified into four separate chemical classes: alkaloids, flavonoids, phenols, and terpenes. Although the focus has been on the cannabinoid-type constituents of cannabis, the non-cannabinoids are just as vital and essential to define the chemical profile and the characteristic properties associated with *C. sativa*. This chapter addresses the isolation and identification of the most important and predominant non-cannabinoid-type constituents, classified into four major classes.

Keywords: cannabis, non-cannabinoids, isolation, structural elucidation, volatile oil

4.1 Alkaloids

A total of only two spermidine alkaloids have been identified in cannabis. The first spermidine alkaloid was isolated in 1975 from a methanolic extract of the roots of a Mexican variant of *C. sativa* that was grown in Mississippi. This alkaloid was later identified as cannabisativine (**1**) through the use of X-ray crystallography [1]. Also, in 1975, cannabisativine (**1**) was isolated from the ethanolic extract of dry leaves and small stems of a cannabis strain grown in Thailand. This extract subjected to both acid–base partitioned and subjected to two different chromatographic techniques, namely both thin-layer chromatography and column chromatography. The compound was then crystallized using acetone [2].

A year later, a second spermidine alkaloid, specifically anhydrocannabisativine (**2**) was isolated from a Mexican variety of cannabis that was grown in Mississippi. The alkaloid was isolated from an extract of dry leaves and roots of this variety, which was subjected to acid–base extraction, followed by silica gel column chromatography. The

*Corresponding author: Mahmoud A. ElSohly,** ElSohly Laboratories, Inc., 5 Industrial Park Drive, Oxford, MS 38655, USA; National Center for Natural Products Research, University of Mississippi, MS 38677, USA; Department of Pharmaceutics, School of Pharmacy, University of Mississippi, MS 38677, USA, e-mails: melsohly@eli.com, melsohly@olemiss.edu; https://orcid.org/0000-0002-2019-2001
Waseem Gul, ElSohly Laboratories, Inc., 5 Industrial Park Drive, Oxford, MS 38655, USA
Shahbaz W. Gul, ElSohly Laboratories, Inc., 5 Industrial Park Drive, Oxford, MS 38655, USA; Sally McDonnell Barksdale Honors College, University of Mississippi, MS 38677, USA; School of Pharmacy, University of Mississippi, MS 38677, USA; Ayub Medical College, Mansehra Road, Abbottabad, Khyber Pakhtunkhwa, Pakistan

https://doi.org/10.1515/9783110718362-004

structural elucidation of anhydrocannabisativine (2) was confirmed via spectral data analysis, as well as through the conversion of cannabisativine (1) to anhydrocannabisativine (2) [3]. A year later, the same research group in Mississippi reported the identification of this compound in 15 different cannabis variants through the use of thin-layer chromatography eluted with a mixture of chloroform:acetone:ammonia (1:1:1). The compound was also identified in the roots and leaves of the same Mexican cannabis variety in 1978 [4]. The chemical structures of both spermidine alkaloids are shown in Figure 4.1.

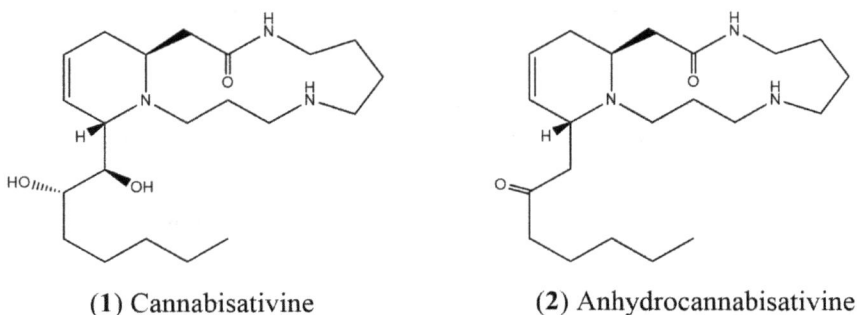

(1) Cannabisativine   (2) Anhydrocannabisativine

Figure 4.1: Chemical structure of cannabis alkaloids.

4.2 Flavonoids

A total of 36 different compounds were isolated from *C. sativa* in the flavonoid category. The basic skeletons of flavonoids can either be geranylated, glycosylated (carbon or oxygen glycosides), methylated, or prenylated. Within the flavonoid class, there are seven basic chemical structures, namely, apigenin, isovitexin, luteolin, kaempferol, orientin, quercetin, and vitexin.

In 1979, three different flavonoid glycosides, namely, vitexin (7), cytisoside (11), and cytisoside glucoside (12), were isolated from the seeds of a Canadian cannabis variety. The chemical structures of these compounds were determined through the use of a hydrolytic test, thin-layer chromatography, and UV spectroscopic analyses [5]. Turner et al. published a review in 1980, having the details of the isolation and chemical structures of 19 different flavonoids (3–6, 8–10, 13–19, 22–23, and 28–30) [6]. Two years later, two methylated isoprenoid flavones, canniflavone 1 and canniflavone 2, were isolated from a *Cannabis* strain grown in Thailand [7]. In 1986, the same two compounds were isolated from an ethanolic extract of cannabis; however, the group renamed the two compounds to cannflavin A (24); a geranyl flavone, and cannflavin B (25); a prenyl flavone. The structures of these compounds were both determined and confirmed through a series of different spectroscopic techniques, such as ^{1}H-NMR, ^{13}C-NMR, and UV spectroscopy [8].

Also, in 2008, three flavonoids were isolated, namely, 6-prenylapigenin (**20**), cannflavin C (**26**), and chrysoeriol (**27**), by Radwan et al. from polar fractions of a cannabis variety grown in Mississippi. Various chromatographic techniques were utilized, namely silica gel column chromatography, reversed-phase high-performance liquid chromatography (HPLC), and vacuum liquid chromatography (VLC). According to both 1D- and 2D-NMR, it was determined that instead of the C-6, the geranyl moiety in cannflavin C (**26**) was attached to C-8 [9]. In 2008, apigenin-6,8-di-C-β-D-glucopyranoside (**21**) was identified in a methanolic extract of cannabis [10]. From cannabis plants of the Mexican variety grown at the University of Mississippi, the pollen grains of the male plants were analyzed and shown to contain two flavonoid glycosides, namely kaempferol-3-*O*-sophoroside (**31**) and quercetin-3-*O*-sophoroside (**32**); their chemical structures were determined through the use of both 1D- and 2D-NMR and UV spectroscopic analyses [11]. In 2012, the ethanolic extract of hemp pectin was analyzed and a flavonoid glycoside, namely rutin (**33**), was identified. The extract went through several purification steps using a variety of techniques, including silica gel column chromatography, macroreticular resin, and Sephadex-LH-20. In addition, for the spectroscopic identification of its chemical structure, a variety of different methodologies were used, including ^{1}H-NMR, ^{13}C-NMR, and electrospray ionization mass spectrometry (ESI-MS) [12]. From an hydroalcoholic extract of hemp inflorescence of monoecious cultivars grown in Italy, three flavonoids, quercetin (**34**), naringenin (**35**), and naringin (**36**) were identified [13]. Using HPLC-PDA, four cultivars were analyzed over time from flowering to ripening. Quercetin (**34**) was present in all four cultivars, while naringenin (**35**) and naringin (**36**) were detected in only two cultivars. From a water extract of industrial hemp of Futura 75 cultivar, which was also cultivated in Italy, naringenin (**35**) was detected, and HPLC-diode array detection-MS was used to quantify it [14]. The chemical structures of these 36 compounds are shown in Figure 4.2.

4.3 Non-cannabinoid phenols

A total of 42 compounds were classified as non-cannabinoid phenols. This class is divided into four different subclasses: dihydrophenanthrenes, dihydrostilbenes, simple phenols, and spiro-indans.

4.4 Dihydrophenanthrenes

From a *Cannabis* strain from Thailand, two different dihydrophenanthrenes were isolated, namely cannabidihydrophenanthrene (cannithrene 1) (**37**) and cannithrene 2 (**38**), reported in 1979 and 1982, respectively [7, 15]. In order to confirm the chemical structure of cannithrene 2 (**38**), X-ray crystallography was used through the analysis of its diacetate derivative [15]. In 2008, using a variety of different normal and reversed-phase chromatography

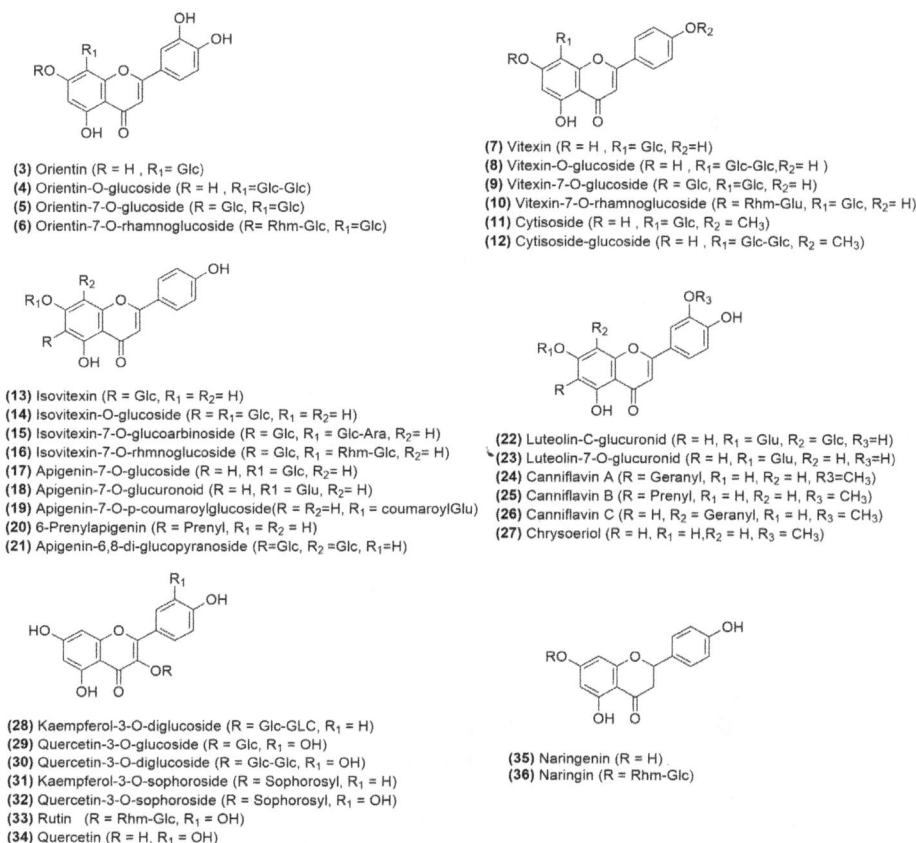

(3) Orientin (R = H , R₁= Glc)
(4) Orientin-O-glucoside (R = H , R₁=Glc-Glc)
(5) Orientin-7-O-glucoside (R = Glc, R₁=Glc)
(6) Orientin-7-O-rhamnoglucoside (R= Rhm-Glc, R₁=Glc)

(7) Vitexin (R = H , R₁= Glc, R₂=H)
(8) Vitexin-O-glucoside (R = H , R₁= Glc-Glc,R₂= H)
(9) Vitexin-7-O-glucoside (R = Glc, R₁=Glc, R₂= H)
(10) Vitexin-7-O-rhamnoglucoside (R = Rhm-Glu, R₁= Glc, R₂= H)
(11) Cytisoside (R = H , R₁= Glc, R₂ = CH₃)
(12) Cytisoside-glucoside (R = H , R₁= Glc-Glc, R₂ = CH₃)

(13) Isovitexin (R = Glc, R₁ = R₂= H)
(14) Isovitexin-O-glucoside (R = R₁= Glc, R₁ = R₂= H)
(15) Isovitexin-7-O-glucoarbinoside (R = Glc, R₁ = Glc-Ara, R₂= H)
(16) Isovitexin-7-O-rhmnoglucoside (R = Glc, R₁ = Rhm-Glc, R₂= H)
(17) Apigenin-7-O-glucoside (R = H, R1 = Glc, R₂= H)
(18) Apigenin-7-O-glucuronoid (R = H, R1 = Glu, R₂= H)
(19) Apigenin-7-O-p-coumaroylglucoside(R = R₂=H, R₁ = coumaroylGlu)
(20) 6-Prenylapigenin (R = Prenyl, R₁ = R₂ = H)
(21) Apigenin-6,8-di-glucopyranoside (R=Glc, R₂ =Glc, R₁=H)

(22) Luteolin-C-glucuronid (R = H, R₁ = Glu, R₂ = Glc, R₃=H)
(23) Luteolin-7-O-glucuronid (R = H, R₁ = Glu, R₂ = H, R₃=H)
(24) Canniflavin A (R = Geranyl, R₁ = H, R₂ = H, R3=CH₃)
(25) Canniflavin B (R = Prenyl, R₁ = H, R₂ = H, R₃ = CH₃)
(26) Canniflavin C (R = H, R₂ = Geranyl, R₁ = H, R₃ = CH₃)
(27) Chrysoeriol (R = H, R₁ = H,R₂ = H, R₃ = CH₃)

(28) Kaempferol-3-O-diglucoside (R = Glc-GLC, R₁ = H)
(29) Quercetin-3-O-glucoside (R = Glc, R₁ = OH)
(30) Quercetin-3-O-diglucoside (R = Glc-Glc, R₁ = OH)
(31) Kaempferol-3-O-sophoroside (R = Sophorosyl, R₁ = H)
(32) Quercetin-3-O-sophoroside (R = Sophorosyl, R₁ = OH)
(33) Rutin   (R = Rhm-Glc, R₁ = OH)
(34) Quercetin (R = H, R₁ = OH)

(35) Naringenin (R = H)
(36) Naringin (R = Rhm-Glc)

Figure 4.2: Chemical structures of cannabis flavonoids.

techniques, three different compounds, specifically two dihydrophenanthrenes (**39, 40**) and one phenanthrene derivative (**41**), were isolated and identified from the ethanolic extract of a *Cannabis* strain in Mississippi. These compounds were then identified to be 4,5-dihydroxy-2,3,7-trimethoxy-9,10-dihydrophenanthrene (**39**), 4-hydroxy-2,3,6,7-tetrame-thoxy-9,10-dihydrophenanthrene (**40**), and 4,7-dimethoxy-1,2,5-trihydroxyphenanthrene (**41**) using IUPAC nomenclature [9]. In the same year, an acetone extract of the *Cannabis* strain (CARMA) was subjected to both fractionation and column chromatography to yield denbinobin (**42**), which was identified and purified through crystallization from ether [16]. Using three different spectroscopic techniques, namely 1D- and 2D-NMR and ESI-MS, 2,3,5,6-tetramethoxy-9,10-dihydrophenanthrenedione (**43**) was isolated from both the leaves and branches of cannabis [17]. Both denbinobin (**42**) and 2,3,5,6-tetramethoxy-9, 10-dihydrophenanthrene-1,4-dione (**43**) are 1,4-phenanthrenequinone derivatives. The chemical structures of these seven non-cannabinoid constituents are shown in Figure 4.3.

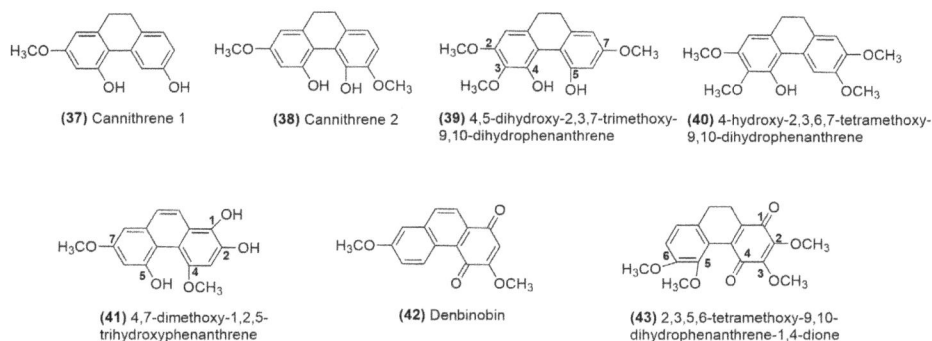

(37) Cannithrene 1

(38) Cannithrene 2

(39) 4,5-dihydroxy-2,3,7-trimethoxy-9,10-dihydrophenanthrene

(40) 4-hydroxy-2,3,6,7-tetramethoxy-9,10-dihydrophenanthrene

(41) 4,7-dimethoxy-1,2,5-trihydroxyphenanthrene

(42) Denbinobin

(43) 2,3,5,6-tetramethoxy-9,10-dihydrophenanthrene-1,4-dione

Figure 4.3: Chemical structure of cannabis dihydrophenanthrene derivatives.

4.5 Dihydrostilbenes

To date, there have been a total of 13 dihydrostilbenes reported in cannabis. In 1979, a series of four different dihydrostilbenes were reported, namely 3-[2-(4-hydroxyphenyl)-ethyl]-5-methoxyphenol (**44**), 3-[2-(3-hydroxy-4-methoxyphenyl)-ethyl]-5-methoxyphenol (**45**), 3-[2-(3-isoprenyl-4-hydroxy-5-methoxy-phenyl)-ethyl]-5-methoxyphenol (**46**), and canniprene (**47**), and their structures were elucidated using both chemical and spectral analyses; however, the chemical structure of canniprene (**47**) was confirmed through synthesis [15]. Five years later, two dihydrostilbenes, specifically cannabistilbene I (**48**) and cannabistilbene II (**49**), were identified from a Panamanian variety of cannabis grown at the University of Mississippi; the chemical structure of cannabistilbene I (**48**) was confirmed through spectroscopic analyses, using a variety of different techniques, including ^{1}H-NMR and mass spectral analysis. In addition, the chemical structure of this compound was confirmed through total synthesis. On the other hand, the chemical structure of cannabistilbene II (**49**) was theorized to have a chemical structure of either **49A** or **49B** [18]. In the same year, another dihydrostilbene, namely 3,4,5-trihydroxydihydrostilbene (**50**), was isolated from an ethanolic extract of a hashish sample, with the structural elucidation of the dihydrostilbene and its triacetate derivative carried out using both ^{1}H-NMR and ^{13}C-NMR spectroscopic techniques and total synthesis [19]. In 2018, three new and two known dihydrostilbenes were identified in a cannabis variety grown in the Yunnan Province of China. These five dihydrostilbenes were tentatively identified as α,α'-dihydro-3′,4,5-trihydroxy-4′-methoxy-3-isopentenylstilbene (**51**), α,α'-dihydro-3,4′,5-trihydroxy-4-methoxy-2,6-diisopentenylstilbene (**52**), and α,α'-dihydro-3′, 4,5-trihydroxy-4′-methoxy-2′,3-diisopentenylstilbene (**53**), α,α'-dihydro-3,4′,5-trihydroxy-4, 5′-diisopentenylstilbene (**54**), and combretastatin B-2 (**55**) through the use of various chromatographic technologies in the isolation and purification of these compounds, including silica gel column chromatography, Sephadex column chromatography, preparative HPLC, and ODS-C18-silica gel column chromatography [20]. The chemical structures of these 13 compounds are shown in Figure 4.4.

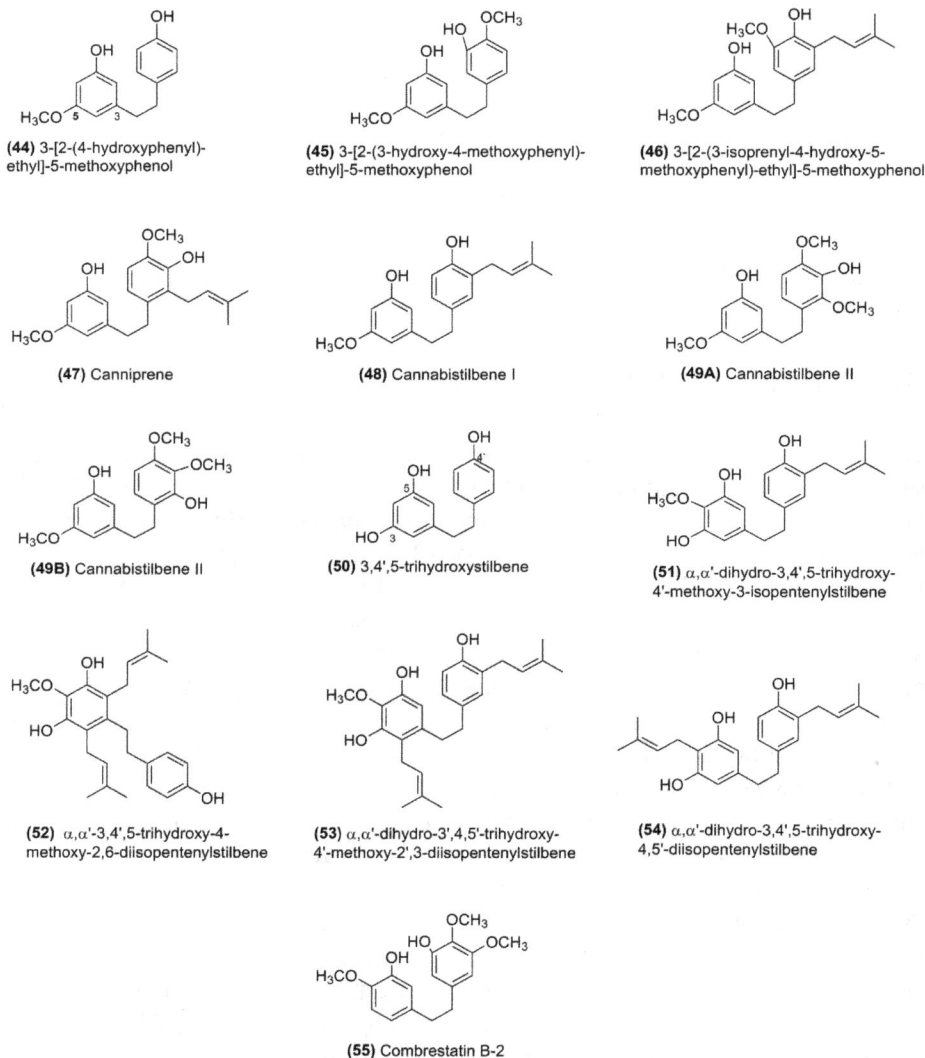

(44) 3-[2-(4-hydroxyphenyl)-ethyl]-5-methoxyphenol

(45) 3-[2-(3-hydroxy-4-methoxyphenyl)-ethyl]-5-methoxyphenol

(46) 3-[2-(3-isoprenyl-4-hydroxy-5-methoxyphenyl)-ethyl]-5-methoxyphenol

(47) Canniprene

(48) Cannabistilbene I

(49A) Cannabistilbene II

(49B) Cannabistilbene II

(50) 3,4',5-trihydroxystilbene

(51) α,α'-dihydro-3,4',5-trihydroxy-4'-methoxy-3-isopentenylstilbene

(52) α,α'-3,4',5-trihydroxy-4-methoxy-2,6-diisopentenylstilbene

(53) α,α'-dihydro-3',4,5'-trihydroxy-4'-methoxy-2',3-diisopentenylstilbene

(54) α,α'-dihydro-3,4',5-trihydroxy-4,5'-diisopentenylstilbene

(55) Combrestatin B-2

Figure 4.4: Chemical structure of cannabis dihydrostilbenes.

4.6 Simple phenols

In the essential oil of cannabis, there were five simple phenols which were detected and identified using gas chromatography-mass spectrometry (GC-MS), namely, eugenol (56), methyleugenol (57), *iso*-eugenol (58), *trans*-anethol (59), and *cis*-anethol (60) [7, 21]. In 1994, the stem exudate of a strain of *C. sativa* underwent acid hydrolysis, and the aglycone of phloroglucinol-β-D-glucoside (61) was isolated; however, phloroglucinol-β-D-glucoside (61) itself was identified via different spectroscopic methodologies, including thin-layer chromatography, ^{1}H-NMR, and GC-MS analyses [22]. Approximately 18 years later, a new simple phenol, namely vanillin (62), was isolated from hemp pectin using silica gel column chromatography; the simple phenol was identified using various spectroscopic technologies, including ^{1}H-NMR, ^{13}C-NMR, and ESI-MS [12]. The chemical structures of all seven simple phenols are shown in Figure 4.5.

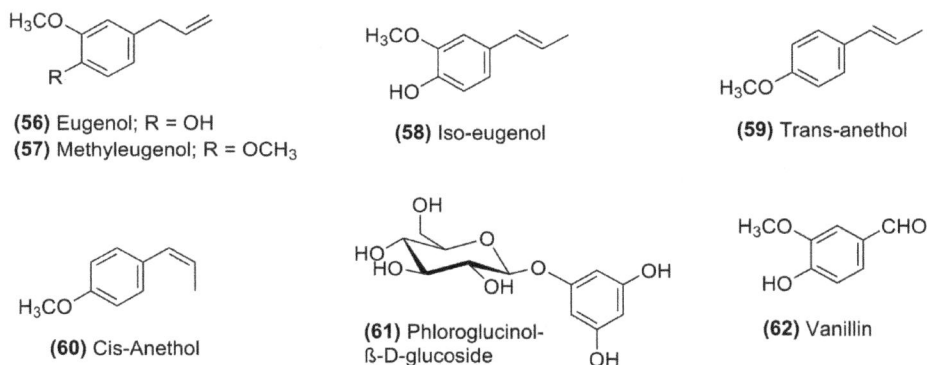

(56) Eugenol; R = OH
(57) Methyleugenol; R = OCH$_3$

(58) Iso-eugenol

(59) Trans-anethol

(60) Cis-Anethol

(61) Phloroglucinol-ß-D-glucoside

(62) Vanillin

Figure 4.5: Chemical structures of cannabis simple phenols.

4.7 Spiro-indans

There are 16 spiro-indan-type compounds isolated from cannabis (63–78, Figure 4.6).

In 1976, cannabispiran (63) was isolated from an Indian *Cannabis* strain using normal-phase silica gel column chromatography, and its structure was confirmed using X-ray crystallography [23]. Cannabispiran (63) was also isolated in the same year from a South African cannabis variety and dubbed the name cannabispirone. The same group also identified cannabispirenone (64) from the same cannabis variety. The chemical structures of both cannabispiran (63) and cannabispirenone (64) were confirmed via NMR and mass spectroscopic analyses [24]. Two years later, an isomer of cannabispirenone (65) was isolated from a cannabis variety from Mexico, with its structure elucidated using both ^{1}H-NMR and EIMS analyses. The isomer itself has both interchangeable hydroxyl and methoxy groups [25]. A new spiro-indan, cannabispiradienone (66), was

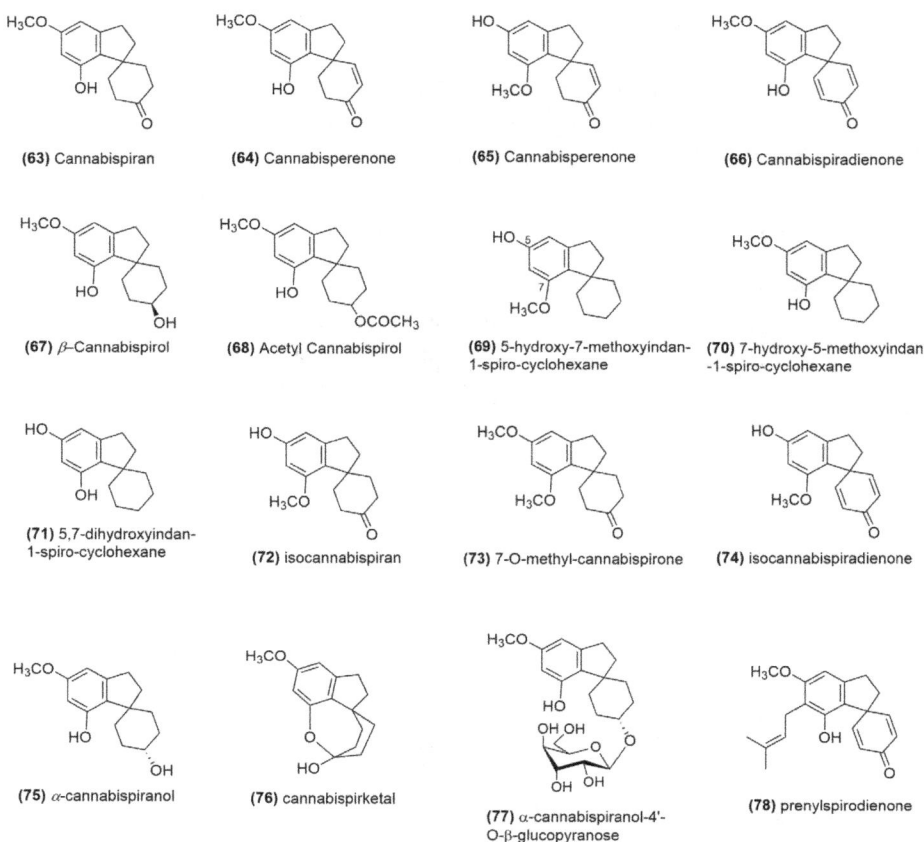

(63) Cannabispiran   (64) Cannabisperenone   (65) Cannabisperenone   (66) Cannabispiradienone

(67) β-Cannabispirol   (68) Acetyl Cannabispirol   (69) 5-hydroxy-7-methoxyindan-1-spiro-cyclohexane   (70) 7-hydroxy-5-methoxyindan-1-spiro-cyclohexane

(71) 5,7-dihydroxyindan-1-spiro-cyclohexane   (72) isocannabispiran   (73) 7-O-methyl-cannabispirone   (74) isocannabispiradienone

(75) α-cannabispiranol   (76) cannabispirketal   (77) α-cannabispiranol-4'-O-β-glucopyranose   (78) prenylspirodienone

Figure 4.6: Chemical structure of cannabis spiro-indans.

identified and isolated from a strain of cannabis grown in Thailand; its chemical structure was determined via ^{1}H-NMR spectroscopic analysis and confirmed using hydrogenation [26]. The hydrogenation of cannabispiradienone (**66**) yielded cannabispiran (**63**), confirming its structure. In addition, through a diene–phenol rearrangement, cannabispiradienone (**66**) can be biogenetically altered to yield cannithrene 1 (**37**) [15]. In 1977, cannabispirol (**67**) was isolated and identified from a variant of South African cannabis grown at the University of Mississippi; the orientation of the hydroxyl group was determined (and therefore better named as β-cannabispirol) through its ^{1}H-NMR analysis (**67**) and that of its acetate derivative, namely acetyl cannabispirol (**68**) [27]. A year later, benzene extract of the dried leaves of Japanese cannabis was chromatographed using a polyamide column followed by silica gel column chromatography, yielding both cannabispirol (**67**) and acetyl cannabispirol (**68**) [15]. A seized hashish sample from Saudi Arabia was analyzed through a series of different spectral and chemical analyses, including flash chromatography and silica gel column chromatography, yielding three spiro-indans which were chemically identified as 5-hydroxy-7-methoxyindan-1-spiro-cyclohexane (**69**), 7-hydroxy-5-methoxyindan-1-spiro-cyclohexane (**70**), and 5,7-dihydroxyindan-1-spiro-cyclo-hexane (**71**) [18]. Through the use of repetitive chromatography of an extract of a canna-bis variety from Panama isocannabispiran (**72**) was identified; the chemical structure of which was determined to be 5′-hydroxy-7′-methoxy-spiro-(cyclohexane-1,1′-indan)-4-one through spectroscopic analysis and structural comparison to cannabispiran (**63**) [28]. In 2008, a new spiro-indan, namely 7-*O*-methyl-cannabispirone (**73**), was isolated from a high-potency cannabis strain using both normal-phase column chromatography and C18-HPLC [29]. Approximately 3 years later, the DCM extract of decarboxylated *C. sativa* was analyzed using C18-flash column chromatography, silica gel column chromatography, and HPLC, and two new compounds were identified, namely isocannabispiradienone (**74**) and *α*-cannabispiranol (**75**). The chemical structures of these compounds were deter-mined through the use of multiple spectroscopic techniques, namely, HR-ESI-MS, ^{1}H-NMR, ^{13}C-NMR, HSQC (heteronuclear single quantum coherence spectroscopy), and heteronu-clear multiple bond correlation (HMBC) [30]. Three new cannabispirans were identified from an ethanolic extract of the leaves of cannabis, with their chemical structures deter-mined through a series of different spectroscopic techniques, including ^{1}H-NMR, ^{13}C-NMR, correlation spectroscopy, HSQC, HMBC, and rotating frame Overhauser enhancement spectroscopy. These compounds were cannabispirketal (**76**), *α*-cannabispiranol-4′-*O*-*β*-glucopyranose (**77**), and prenylspirodienone (**78**) [31]. A year later, the chemical structure of prenylspirodienone (**78**) was proved through the use of extensive NMR and ESI-MS analyses. In addition, the biosynthetic pathway for this compound was proposed by the group [32]. The structures of these 16 compounds are shown in Figure 4.6.

4.8 Terpenes

Terpenes represent the second largest class of cannabis constituents. They are primar-ily responsible for the characteristic smell of the cannabis plant. Terpenes can pro-duce cannabimimetic activity both *in vivo* and *in vitro* which may provide a decent evidence for the entourage effect of cannabis extract along with cannabinoids [33].

Several groups have reviewed the total number of terpenes identified in cannabis [34–37], and a total of 120 or more terpenes were proposed. However, there are actu-ally a total of 117 terpenes. This discrepancy arose from the misnaming of 3 different terpenes each with 2 different names, thus each was counted twice in the 120 count.

The terpenes are known to be acyclic, monocyclic, and polycyclic hydrocarbons, with substituted groups including alcohols, aldehydes, ketones, and esters [36]. This class is subdivided into 5 different groups, namely, 61 monoterpenes, 51 sesquiter-penes, 2 diterpenes, 2 triterpenes, and 4 miscellaneous compounds. Monoterpenes have a C_{10} backbone, sesquiterpenes have a C_{15} backbone, diterpenes have a C_{20}

backbone, and triterpenes have a C_{30} backbone. Two sesquiterpenes and 1 diterpene were recently isolated in 2020 and 2021, making a total of 120 terpenes.

As was stated by Fischedick, cannabis cultivars with similar cannabinoid content can be distinguished by their terpene content, especially in high tetrahydrocannabinol cultivars using GC-FID (flame ionization detection) [38]. In addition, cannabis terpene profiles are strongly related to the diversity of terpene synthase (TPS) enzymes found in cannabis TPS gene family [39]. Extracts of cannabis inflorescence were analyzed by GC/FID in order to get the chemotaxonomic discrimination between different cannabis biotypes [40]. Using GC/FID analysis, a metabolomic approach has been adopted in order to fill the gap between scientists and patients. A detailed terpene profile has been reported by the analysis of 44 major terpenes which led to identify cannabis markers and differentiate between *C. sativa* and *C. indica* [41]. Eleven cannabis varieties have been chemically discriminated based on the analysis of their cannabinoid and terpene contents followed by principal component analysis [42]. Chemical composition of medicinal cannabis oils including cannabinoids and terpenes has been reported in order to better understand the experience of the self-medicating patients [43].

4.9 Monoterpenes

A total of 61 monoterpenes have been reported. There are two separate classes into which these compounds can be classified into oxygenated monoterpenes and monoterpene hydrocarbons. These two classifications can further be broken down into three different types: acyclic (linear), monocyclic, or bicyclic (79–97) and (98–139). The chemical structures of cannabis monoterpenes (hydrocarbon derivatives) are shown in Figure 4.7, while chemical structures of cannabis oxygenated monoterpenes (acyclic, monocyclic, and bicyclic) are shown in Figure 4.8.

The first report of monoterpenes was in 1942, when two new monoterpene hydrocarbons, namely *p*-cymene (82) and minor quantities of 1-methyl-4-isoprenyl-benzene or dehydro-*p*-cymene (83), from the analysis of the low-boiling-point terpene fraction of Egyptian hashish were reported [44]. In approximately 1962, two other monoterpenes, namely myrcene (79), an acyclic monoterpene, and a monocyclic monoterpene, limonene (89), were both isolated and identified in the essential oil of wild *C. sativa* from Canada [2].

Through hydrodistillation of fresh, Indian *C. sativa*, the essential oil of the sample was obtained, from which various terpenoid compounds were identified. After fractional distillation of the essential oil, five different fractions were generated; the fifth fraction was further purified through the use of column chromatography with alumina as the stationary phase and petroleum ether, benzene, ether, and alcohol successively as the mobile phase. The fractions generated from the use of petroleum ether were combined and nominally identified as Fraction 5A, and the fractions generated from

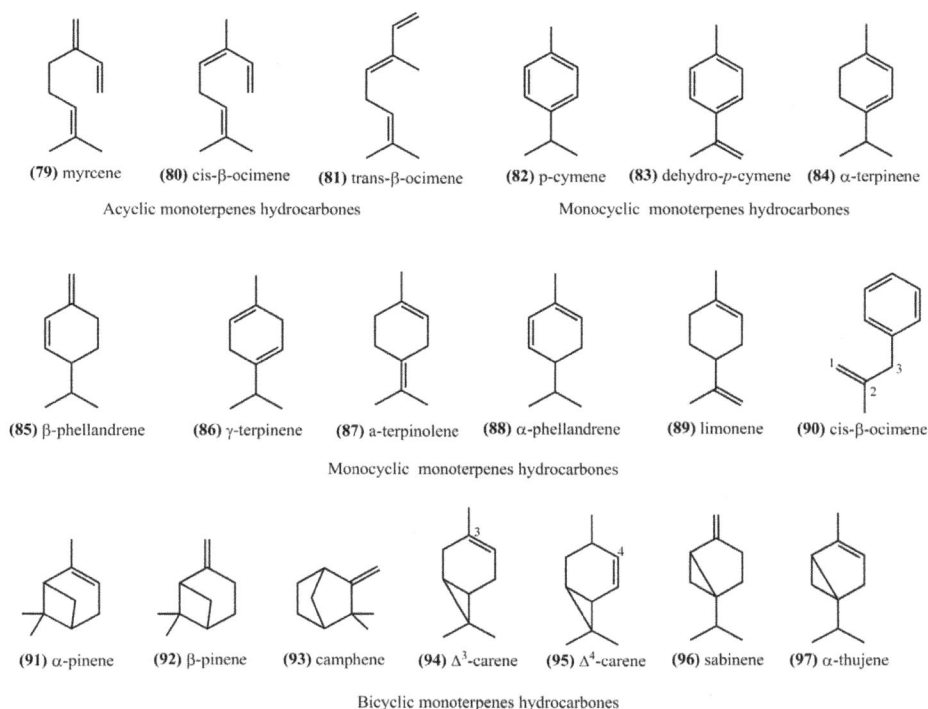

Figure 4.7: Chemical structure of cannabis monoterpenes (hydrocarbon derivatives).

the use of benzene were combined and identified as Fraction 5B. A total of 24 different terpenoid-type compounds were detected, from which 12 monoterpenes were previously unreported, namely α-terpinene (**84**), β-phellandrene (**85**), γ-terpinene (**86**), α-terpinolene (**87**), α-pinene (**91**), β-pinene (**92**), camphene (**93**), linalool (**98**), α-terpineol (**109**), terpinene-4-ol (**110**), linalool oxide (**120**), and sabinene hydrate (**126**) [45].

Both Dutch and Turkish cannabis volatile oil samples were analyzed in 1971 and 1973 through the use of capillary GC. These volatile oils were obtained through the use of two different methods, namely nitrogen extraction and hydrodistillation. There was a total of 18 monoterpenes identified from these analyses, from which seven were not reported previously, namely *cis*-β-ocimene (**80**), *trans*-β-ocimene (**81**), α-phellandrene (**88**), Δ^3-carene (**94**), Δ^4-carene (**95**), sabinene (**96**), and α-thujene (**97**) [46, 47]. Three years later, a new oxygenated monoterpene, namely, *m*-mentha-1,8-(9)-dien-5-ol (**106**) was isolated from the volatile oil of cannabis [48].

One of the most characteristic properties of *C. sativa* is its unique aroma, as this is one of the first modes of detection for illicit marijuana trafficking. In order to analyze the headspace components of this plant, a direct gas chromatographic analysis was conducted. The real samples used in this study were obtained from customs seizures while the standard for this study was grown and harvested at the University of

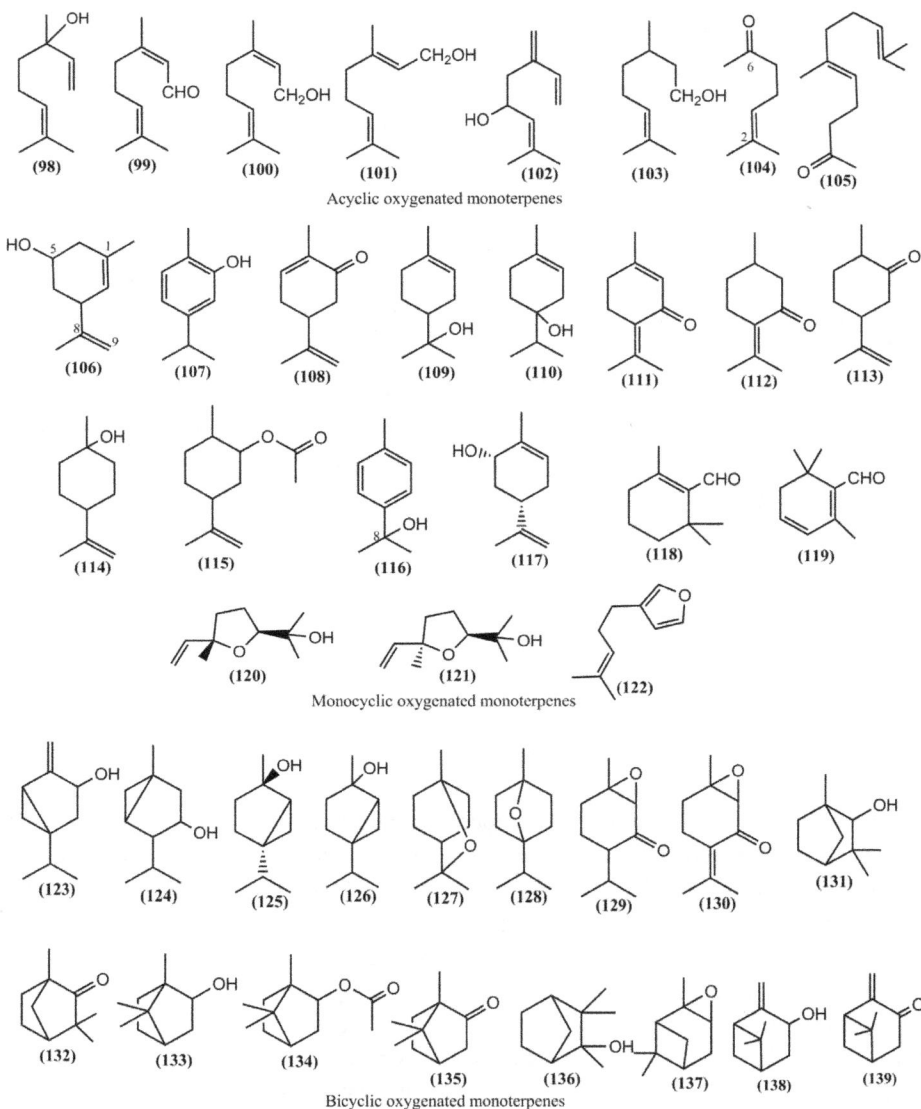

Figure 4.8: Chemical structure of cannabis oxygenated monoterpenes (acyclic, monocyclic, and bicyclic).

Mississippi. Approximately 1 g of each sample was weighed, placed in a microvial, and then heated at 65 °C for approximately 1 h. Using a gas-tight syringe, 5 mL of headspace air was taken and directly injected into the gas chromatograph. From this analysis, 18 different terpenoid-type compounds were detected, from which 3 compounds were not previously reported: 2-methyl-2-heptene-6-one (**104**), fenchyl alcohol (**131**), and borneol (**133**) [49].

The volatile oil from cannabis grown and harvested in Mexico was prepared and analyzed by GC-MS. A total of 17 monoterpenes were identified, from which 6 oxygenated monoterpenes were not reported previously, namely nerol (**100**), geraniol (**101**), carvacrol (**107**), 1,8-cineol (**127**), 1,4-cineol (**128**), and camphor (**135**) [48]. Also, in 1974, the GC-MS analysis and retention time matching were carried out for the volatile oil of *C. sativa*, and a new monoterpene, namely piperitenone (**111**), was detected and identified [50].

In both 1975 and 1978, a total of 55 different monoterpene components were reported, from which 24 compounds were reported for the first time, namely 3-phenyl-2-methyl-prop-1-ene (**90**), and 23 oxygenated hydrocarbons, namely citral B (**99**), citronellol (**103**), geranyl acetone (**105**), carvone (**108**), pulegone (**112**), dihydrocarvone (**113**), β-terpineol (**114**), dihydrocarveyl acetate (**115**), *p*-cymene-8-ol (**116**), β-cyclocitral (**118**), safranal (**119**), *cis*-linalool oxide (**121**), perillene (**122**), sabinol (**123**), thujyl alcohol (**124**), piperitone oxide (**129**), piperitenone oxide (**130**), fenchone (**132**), bornyl acetate (**134**), camphene hydrate (**136**), α-pinene oxide (**137**), pinocarveol (**138**), and pinocarvone (**139**) [20, 51].

Through collecting the cannabis essential oil prepared by the steam distillation of marijuana buds using the lighter than water volatile oil apparatus, it was possible to study the composition of the oil itself. The samples investigated in this study were dried and stored at room temperature for three different time points: 1 week, 1 month, and 3 months. These oils were analyzed using both GC-MS and GC-FID methods; the percentage composition of oils was determined through the use of GC-FID, and the detection of terpenes was identified through the use of GC-MS. From the analyses, there were a total of 68 compounds identified in the study, with 57 of the components being classified as terpenoid-type components and other compounds like ketones and esters. The fresh buds' oil was analyzed and identified to contain the greatest composition of monoterpenes, with a calculated value of approximately 92.48%. From this analysis, it was determined that 6.84% of the terpenes identified were sesquiterpenes, and the other components were approximately 0.68%. The 3-month samples were analyzed to identify differences between the three time points. From this analysis, it was determined that there was an increase in the percent composition of sesquiterpenes (35.63%) and other compounds (2.35%) with a proportional decrease in the percent composition of monoterpenes (62.02%) of the volatile oil. From this study, there were a total of three new oxygenated monoterpenes that were reported, namely ipsdienol (**102**), *cis*-carveol (**117**), and *cis*-sabinene hydrate (**125**) [52].

4.10 Sesquiterpenes

The first sesquiterpene to be identified, namely, α-caryophyllene (α-humulene) (**140**) was reported in a higher boiling point fraction of Egyptian hashish in 1942 [44]. Through GC analysis of the volatile oil of fresh cannabis, a new sesquiterpene, namely, β-caryophyllene (**141**) was detected in 1961 [53]. Four years later, the volatile oil of cannabis grown in India was analyzed and five new sesquiterpenes were identified: caryophyllene oxide (**142**), curcumene (**143**), α-*trans*-bergamotene (**144**), α-selinene (**145**), and β-farnesene (**146**) [45]. In 1973, the volatile oil of cannabis was analyzed and four new sesquiterpenes were detected and reported, namely longifolene (**147**), humulene epoxide I (**148**), humulene epoxide II (**149**), and caryophyllene alcohol (caryophyllenol) (**150**) [54]. The sesquiterpene β-bisabolene (**151**) was detected in only one study, where the volatile oil, headspace volatiles, and samples of customs seizures of marijuana were analyzed [55]. The next year, the essential oil of Mexican cannabis was analyzed through GC-FID and GC-MS analysis, and three sesquiterpenes were reported, namely allo-aromadendrene (**152**), calamenene (**153**), and α-copaene (**154**) [56].

In 1974, the essential oil of cannabis from Mexico was analyzed and a new sesquiterpene was identified for the first time, namely, nerolidol (**155**), in addition to 17 different monoterpenes [44]. In addition, using both GC-MS and GC analyses, a new sesquiterpene, namely α-gurjunene (**156**), was identified in cannabis resin [56].

There were four sesquiterpenes, namely, iso-caryophyllene (**157**), β-selinene (**158**), selina-3,7(11)-diene (**159**), and selina-4(14),7(11)-diene (**160**) that were identified from the analysis of the essential oil of cannabis [51]. These sesquiterpenes were confirmed 3 years later by the same group through the use of both GC and GC-MS analyses [21].

In 1978, α-gurjunene (**159**) was confirmed to be present in cannabis; however, there were additional sesquiterpenes that were reported for the first time, namely α-bisabolol (**161**), α-cedrene (**162**), α-cubebene (**163**), δ-cadinene (**164**), epi-β-santalene (**165**), farnesol (**166**), γ- cadinene (**167**), γ-elemene (**168**), γ-eudesmol (**169**), guaiol (**170**), ledol (**171**), *trans-trans*-α-farnesene (**172**), (Z)-β-farnesene (**173**), and farnesyl acetone (**174**) [21].

There were 14 new sesquiterpenes which were isolated and identified in 1996, namely, α-cadinene (**175**), α-*cis*-bergamotene (**176**), α-eudesmol (**177**), α-guaiene (**178**), α-longipinene (**179**), α-ylangene (**180**), β-elemene (**181**), β-eudesmol (**182**), epi-α-bisabolol (**183**), γ-*cis*-bisabolene (**184**), γ-curcumene (**185**), γ-muurolene (**186**), γ-*trans*-bisabolene (**187**), and viridiflorene (**188**) [51].

Germacrene-B (**189**) was detected and quantified by GC-MS for the first time from the hemp essential oil, while from the organic extract of cannabis inflorescence of Ferimon and Uso-31 cultivars, clovandiol (**190**) was identified [13, 57]. The structures of these sesquiterpenes are shown in Figure 4.9.

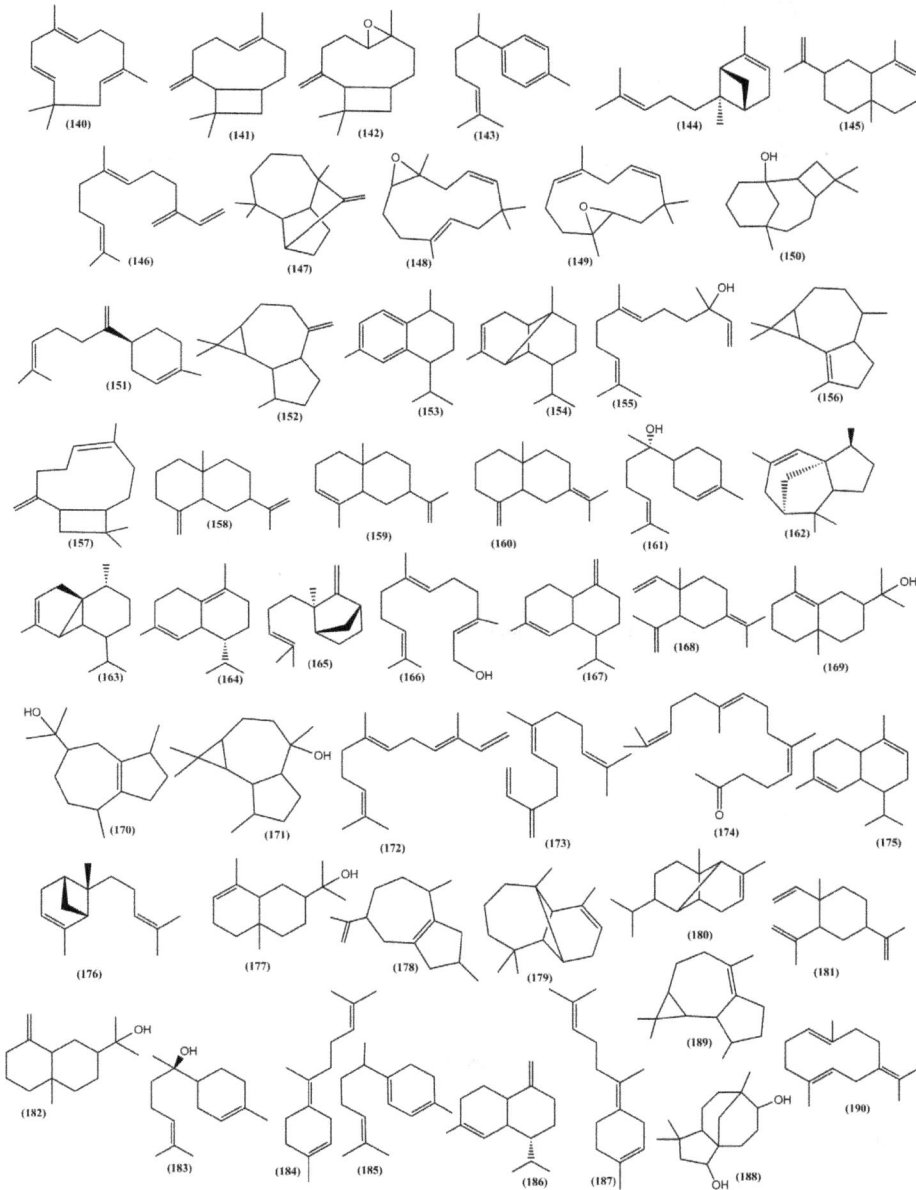

Figure 4.9: Chemical structures of cannabis sesquiterpenes.

4.11 Diterpenes

There have been only two diterpenes, namely, phytol (**191**) and neophytadiene (**192**), which have been isolated from the volatile oil of cannabis. Column chromatography was used to purify the oil, and GC-MS was used to identify the diterpenes in the oxygenated compound fraction [13, 21]. The structures of these compounds can be found in Figure 4.10.

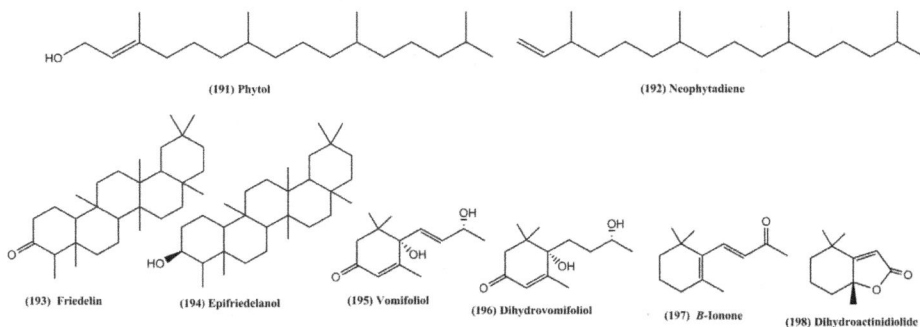

Figure 4.10: Chemical structures of cannabis diterpenes, triterpenes, and miscellaneous terpenes.

4.12 Triterpenes

There have been two different triterpenes which have been isolated from *C. sativa*. The ethanolic extract of cannabis roots was analyzed, and two different triterpenes were identified, namely, friedelin (friedelan-3-one, **193**) and epifriedelanol (**194**). These compounds were confirmed using spectral analysis and comparison with authentic samples [58]. The structures of these two triterpenes are shown in Figure 4.10.

4.13 Miscellaneous terpenes

To date, there have been only four miscellaneous terpenes which have been isolated from cannabis. In 1978, stems and leaves of cannabis were analyzed for their constituents; using spectra data and total synthesis from (+)-α-ionone, two isophorone-type compounds, namely, vomifoliol (**195**) and dihydrovomifoliol (**196**), were reported [21]. In addition, there were two additional miscellaneous-type compounds, namely, β-ionone (**197**) and dihydroactinidiolide (**198**), which were isolated from the volatile oil of cannabis. The structures of these compounds are shown in Figure 4.10.

4.14 Conclusion

Over 500 different constituents of *C. sativa* have been reported, and the majority of these constituents fall under the non-cannabinoid-type compounds. Although these compounds are not as well investigated as the cannabinoids, these compounds are vital to the function of the plant and contribute to the specific chemical profiles associated with different varieties of the plant. This chapter covered the most important non-cannabinoid constituents, namely, the flavonoids, alkaloids, terpenes, and non-cannabinoid phenols. Due to the passing of the Farm Bill of 2018, there is a renewed interest in the chemical and biological activities of *C. sativa*, resulting in a proportional increase in efforts to investigate activities associated not only with the cannabinoid constituents but also with the non-cannabinoid constituents of the cannabis plant.

References

[1]   Latter HL, Abraham DJ, Turner CE, Knapp JE, Schiff PL, Jr., Slatkin DJ. Cannabisativine, a new alkaloid from *Cannabis sativa* L. root. Tetrahedron Letters 1975; 16: 2815–2818.
[2]   El-Feraly F, Elsohly M, Boeren E, Turner C, Ottersen T, Aasen A. Crystal and molecular structure of cannabispiran and its correlation to dehydrocannabispiran: Two novel cannabis constituents. Tetrahedron 1977; 33: 2373–2378.
[3]   Elsohly M, Turner C. Screening of Cannabis grown from seed of various geographical origins for the alkaloids hordenine, cannabisativine and anhydrocannabisativine. UN Secretariat ST/SOA/SER S/54 1977.
[4]   Elsohly MA, Turner CE, Phoebe CH, Jr., Knapp JE, Schiff PL, Jr., Slatkin DJ. Anhydrocannabisativine, a new alkaloid from *Cannabis sativa* L. Journal of Pharmaceutical Sciences 1978; 67: 124–124.
[5]   Clark MN, Bohm BA. Flavonoid variation in Cannabis L. Botanical Journal of the Linnean Society 1979; 79: 249–257.
[6]   Turner CE, Elsohly MA, Boeren EG. Constituents of *Cannabis sativa* L. XVII. A review of the natural constituents. Journal of Natural Products 1980; 43: 169–234.
[7]   Crombie L, Crombie WML. Natural products of Thailand high Δ 1-THC-strain Cannabis. The bibenzyl-spiran-dihydrophenanthrene group: Relations with cannabinoids and canniflavones. Journal of the Chemical Society, Perkin Transactions 1982; 1: 1455–1466.
[8]   Barrett M, Scutt A, Evans F. Cannflavin A and B, prenylated flavones from *Cannabis sativa* L. Experientia 1986; 42: 452–453.
[9]   Radwan MM, ElSohly MA, Slade D, Ahmed SA, Wilson L, El-Alfy AT, Khan IA, Ross SA. Non-cannabinoid constituents from a high potency *Cannabis sativa* variety. Phytochemistry 2008; 69: 2627–2633.
[10]   Cheng L, Kong D, Hu G. Study on Hemp. I. Chemical constituents from petroleum ether and n-butanol portions of the methanol extract. Chinese Journal of Pharmaceuticals 2008; 39: 18.
[11]   Ross SA, ElSohly MA, Sultana GN, Mehmedic Z, Hossain CF, Chandra S. Flavonoid glycosides and cannabinoids from the pollen of *Cannabis sativa* L. Phytochemical Analysis: An International Journal of Plant Chemical and Biochemical Techniques 2005; 16: 45–48.

[12] Chen B, Cai G, Yuan Y, Li T, He Q, He J. Chemical constituents in hemp pectin I. Chinese Journal of Experimental Traditional Medical Formulae 2012; 18: 98–100.

[13] Ingallina C, Sobolev AP, Circi S, Spano M, Fraschetti C, Filippi A, Di Sotto A, Di Giacomo S, Mazzoccanti G, Gasparrini F. *Cannabis sativa* L. inflorescences from monoecious cultivars grown in central Italy: An untargeted chemical characterization from early flowering to ripening. Molecules 2020; 25: 1908.

[14] Di Giacomo V, Recinella L, Chiavaroli A, Orlando G, Cataldi A, Rapino M, Di Valerio V, Politi M, Antolini MD, Acquaviva A. Metabolomic profile and antioxidant/anti-inflammatory effects of industrial hemp water extract in fibroblasts, keratinocytes and isolated mouse skin specimens. Antioxidants 2021; 10: 44.

[15] Shoyama Y, Nishioka I. Cannabis. XIII. Two new spiro-compounds, cannabispirol and acetyl cannabispirol. Chemical and Pharmaceutical Bulletin 1978; 26: 3641–3646.

[16] Sánchez-Duffhues G, Calzado MA, de Vinuesa AG, Caballero FJ, Ech-Chahad A, Appendino G, Krohn K, Fiebich BL, Muñoz E. Denbinobin, a naturally occurring 1, 4-phenanthrenequinone, inhibits HIV-1 replication through an NF-κB-dependent pathway. Biochemical Pharmacology 2008; 76: 1240–1250.

[17] Cheng L, Kong D, Hu G, Li H. A new 9, 10-dihydrophenanthrenedione from *Cannabis sativa*. Chemistry of Natural Compounds 2010; 46: 710–712.

[18] ElSohly HN, Ma G-E, Turner CE, ElSohly MA. Constituents of *Cannabis sativa*, XXV. Isolation of two new dihydrostilbenes from a Panamanian variant. Journal of Natural Products 1984; 47: 445–452.

[19] El-Feraly FS. Isolation, characterization, and synthesis of 3, 5, 4′-trihydroxybibenzyl from *Cannabis sativa*. Journal of Natural Products 1984; 47: 89–92.

[20] Guo T, Liu Q, Hou P, Li F, Guo S, Song W, Zhang H, Liu X, Zhang S, Zhang J. Stilbenoids and cannabinoids from the leaves of *Cannabis sativa f. sativa* with potential reverse cholesterol transport activity. Food & Function 2018; 9: 6608–6617.

[21] Hendriks H, Malingre TM, Batterman S, Bos R. Essential oil of *cannabis-sativa l*. In: Planta Medica. Georg Thieme Verlag, Po Box 30 11 20, D-70451, Stuttgart, Germany, 1978; 280–281.

[22] Hammond CT, Mahlberg PG. Phloroglucinol glucoside as a natural constituent of *Cannabis sativa*. Phytochemistry 1994; 37: 755–756.

[23] Ottersen T, Aasen A, El-Feraly FS, Turner CE. X-ray structure of cannabispiran: A novel cannabis constituent. Journal of the Chemical Society, Chemical Communications 1976; 15: 580–581.

[24] Bercht C, Van Dongen J, Heerma W, Lousberg RC, Küppers F. Cannabispirone and cannabispirenone, two naturally occurring spiro-compounds. Tetrahedron 1976; 32: 2939–2943.

[25] Kettenes-van den Bosch J, Salemink C. Cannabis XIX. Oxygenated 1, 2-diphenylethanes from Marihuana. Recueil Des Travaux Chimiques Des Pays-Bas 1978; 97: 221–222.

[26] Cromble L, Mary W, Crombie L, Jamieson SV. Isolation of cannabispiradienone and cannabidihydrophenanthrene. Biosynthetic relationships between the spirans and dihydrostilbenes of Thailand Cannabis. Tetrahedron Letters 1979; 20: 661–664.

[27] Boeren E, Elsohly M, Turner C, Salemink C. β-Cannabispiranol: A new non-cannabinoid phenol from *cannabis sativa* L. Experientia 1977; 33: 848–848.

[28] ElSohly HN, Turner C. Iso-cannabispiran, a new spiro compound isolated from Panamenian variant of *Cannabis sativa* L. Experientia 1982; 38: 229–229.

[29] Radwan MM, Ross SA, Slade D, Ahmed SA, Zulfiqar F, ElSohly MA. Isolation and characterization of new cannabis constituents from a high potency variety. Planta Medica 2008; 74: 267–272.

[30] Pagani A, Scala F, Chianese G, Grassi G, Appendino G, Taglialatela-Scafati O. Cannabioxepane, a novel tetracyclic cannabinoid from hemp, *Cannabis sativa* L. Tetrahedron 2011; 67: 3369–3373.

[31] Guo -T-T, Zhang J-C, Zhang H, Liu Q-C, Zhao Y, Hou Y-F, Bai L, Zhang L, Liu X-Q, Liu X-Y. Bioactive spirans and other constituents from the leaves of *Cannabis sativa f. sativa*. Journal of Asian Natural Products Research 2017; 19: 793–802.

[32] Nalli Y, Arora P, Riyaz-Ul-Hassan S, Ali A. Chemical investigation of *Cannabis sativa* leading to the discovery of a prenylspirodinone with anti-microbial potential. Tetrahedron Letters 2018; 59: 2470–2472.

[33] LaVigne JE, Hecksel R, Keresztes A, Streicher JM. *Cannabis sativa* terpenes are cannabimimetic and selectively enhance cannabinoid activity. Scientific Reports 2021; 11: 1–15.

[34] ElSohly MA, Slade D. Chemical constituents of marijuana: The complex mixture of natural cannabinoids. Life Sciences 2005; 78: 539–548.

[35] Flores-Sanchez IJ, Verpoorte R. Secondary metabolism in cannabis. Phytochemistry Reviews 2008; 7: 615–639.

[36] Brenneisen R. Chemistry and analysis of phytocannabinoids and other Cannabis constituents. In: Marijuana and the cannabinoids. In: ElSohly, M.A. (eds), Totowa, New Jersey, Humana Press, 2007; 17–49.

[37] Russo EB. Taming THC: Potential cannabis synergy and phytocannabinoid-terpenoid entourage effects. British Journal of Pharmacology 2011; 163: 1344–1364.

[38] Fischedick JT. Identification of terpenoid chemotypes among high (−)-trans-Δ9-tetrahydrocannabinol -producing *Cannabis sativa* L. cultivars. Cannabis and Cannabinoid Research 2017; 2: 34–47.

[39] Booth JK, Bohlmann J. Terpenes in *Cannabis sativa*–From plant genome to humans. Plant Science 2019; 284: 67–72.

[40] Hillig KW. A chemotaxonomic analysis of terpenoid variation in Cannabis. Biochemical Systematics and Ecology 2004; 32: 875–891.

[41] Hazekamp A, Fischedick J. Cannabis-from cultivar to chemovar. Drug Testing and Analysis 2012; 4: 660–667.

[42] Fischedick JT, Hazekamp A, Erkelens T, Choi YH, Verpoorte R. Metabolic fingerprinting of *Cannabis sativa* L., cannabinoids and terpenoids for chemotaxonomic and drug standardization purposes. Phytochemistry 2010; 71: 2058–2073.

[43] Romano LL, Hazekamp A. Cannabis oil: Chemical evaluation of an upcoming cannabis-based medicine. Cannabinoids 2013; 1: 1–11.

[44] Simonsen JL, Todd AR. 32. Cannabis indica. Part X. The essential oil from Egyptian hashish. Journal of the Chemical Society (Resumed) 1942: 188–191.

[45] Nigam M, Handa K, Nigam I, Levi L. Essential Oils And Their Constituents: XXIX. The essential oil of Marihuana: Composition of genuine Indian *Cannabis Sativa* L. Canadian Journal of Chemistry 1965; 43: 3372–3376.

[46] Bercht C, Kuppers F, Lousberg R. Volatile Constituents of *Cannabis sativa* L.; UN Secretariat Document. In: ST/SUA/SER. 5/29, 22 July 1971; 1971.

[47] Lousberg RJ, Salemink CA. Some aspects of Cannabis research. Pharmaceutisch Weekblad 1973; 108: 1–9.

[48] Radwan MM, Chandra S, Gul S, ElSohly MA. Cannabinoids, phenolics, terpenes and alkaloids of cannabis. Molecules 2021; 26: 2774.

[49] Hood L, Dames M, Barry G. Headspace volatiles of marijuana. Nature 1973; 242: 402–403.

[50] Strömberg L. Minor components of cannabis resin: IV. Mass spectrometric data and gas chromatographic retention times of terpenic components with retention times shorter than that of cannabidiol. Journal of Chromatography A 1974; 96: 99–114.

[51] Hendriks H, Malingré TM, Batterman S, Bos R. Mono-terpene and sesqui-terpene hydrocarbons of essential oil of *Cannabis sativa*. Phytochemistry 1975; 14: 814–815.

[52] Ross SA, ElSohly MA. The volatile oil composition of fresh and air-dried buds of *Cannabis sativa*. Journal of Natural Products 1996; 59: 49–51.

[53] Martin L, Smith DM, Farmilo C. Essential oil from fresh *Cannabis sativa* and its use in identification. Nature 1961; 191: 774–776.

[54] Stahl E, Kunde R. Neue inhaltsstoffe aus dem ätherischen öl von *Cannabis sativa*. Tetrahedron Letters 1973; 14: 2841–2844.

[55] Hood LVS, Dames ME, Barry GT. Headspace volatiles of marijuana. Nature 1973; 242: 402–403. doi: 10.1038/242402a0.

[56] Bercht C, Paris M. Oil of *Cannabis sativa*. Bulletin technique Gattefossé. Bull Tech Gattefosse Sfpa 1973; 68: 87–90.

[57] Menghini L, Ferrante C, Carradori S, D'Antonio M, Orlando G, Cairone F, Cesa S, Filippi A, Fraschetti C, Zengin G. Chemical and bioinformatics analyses of the anti-leishmanial and anti-oxidant activities of hemp essential oil. Biomolecules 2021; 11: 272.

[58] Slatkin DJ, Doorenbos NJ, Harris LS, Masoud AN, Quimby MW, Schiff PL. Chemical constituents of *Cannabis sativa* L. root. Journal of Pharmaceutical Sciences 1971; 60: 1891–1892.

Waseem Gul, Elsayed A. Ibrahim, Shahbaz W. Gul
and Mahmoud A. ElSohly*

5 Total synthesis of the major phytocannabinoids

Abstract: As presented in the previous chapters, the cannabis plant is rich in a variety of constituents belonging to many different chemical classes. The most important and characteristic class, specific to cannabis, is the cannabinoids. While a few naturally occurring cannabinoids are abundant in the plant materials, especially Δ^9-tetrahydrocannabinol and cannabidiol, other cannabinoids are present in small quantities. Synthesis of cannabinoids, like most other natural products, has been one of the methods used for structure elucidation. However, synthesis is also used to produce large supply of natural products that might not be feasible by isolation from plant material. This chapter is dedicated to methods used to synthesize the major cannabinoids that could be used for product development efforts of these cannabinoids. The synthetic routes offer alternative methods to secure drug supply of these cannabinoids. Only examples of synthetic methods are given, with no intention to provide exhaustive coverage of all literature procedures.

Keywords: cannabis, cannabinoids, history, synthesis

5.1 Introduction

Cannabinoids are terpenophenolic secondary metabolites found in *Cannbis sativa* L. (*C. sativa*) consisting of alkylresorcinol and monoterpene units [1]. Up to now, at least 129 known cannabinoids have been identified in or isolated from cannabis. Furthermore, terpenes, nitrogenous compounds, phenolic substances (including flavonoids), sugars as well as several other chemical classes have been reported in cannabis [2].

---

*__Corresponding author: Mahmoud A. ElSohly,__ ElSohly Laboratories, Inc., 5 Industrial Park Drive, Oxford, MS 38655, USA; National Center for Natural Products Research, University of Mississippi University, MS 38677, USA; Department of Pharmaceutics, School of Pharmacy University of Mississippi, University, MS 38677, USA, e-mails: elsohly@elsohly.com; melsohly@olemiss.edu; https://orcid.org/0000-0002-0019-2001

Waseem Gul, ElSohly Laboratories, Inc., 5 Industrial Park Drive, Oxford, MS 38655, USA
Elsayed A. Ibrahim, National Center for Natural Products Research, University of Mississippi, MS 38677, USA; Department of Pharmaceutical Analytical Chemistry, Faculty of Pharmacy, Suez Canal University, Ismailia 41522, Egypt
Shahbaz W. Gul, ElSohly Laboratories, Inc., 5 Industrial Park Drive, Oxford, MS 38655, USA; Sally McDonnell Barksdale Honors College, University of Mississippi, MS 38677, USA; School of Pharmacy, University of Mississippi, MS 38677, USA; Ayub Medical College, Mansehra Road, Abbottabad, KPK, Pakistan

https://doi.org/10.1515/9783110718362-005

The term "cannabinoids" was first coined by Mechoulam and Gaoni [3] in 1967, restricted in natural occurrence only to *C. sativa* and not to any other plants. This includes the analogs of these compounds as well as their transformation products.

In the early 1960s, illicit marijuana use dramatically increased, with over 36 million American users by 1975. Given that over 10% of the users aged 20–24 were daily users, it is no wonder that society was concerned, leading to the intense sociopolitical controversy surrounding marijuana today. Marijuana has centuries-long history of therapeutic use, as documented by Chinese, Indian, and Middle Eastern cultures. Indications for such therapeutic uses include insomnia, migraine pain, asthma, and loss of appetite. In 1971, Hepler and Frank found the psychoactive constituent Δ^9-tetrahydrocannabinol (Δ^9-THC) to reduce intraocular pressure in glaucoma patients [4]. In 1975, Sallan et al. [5] confirmed that Δ^9-THC also reduces nausea in cancer patients undergoing chemotherapy.

The dibenzopyran and monoterpenoid numbering systems are used to describe cannabinoids, with the former being used by chemical abstracts and the latter being used by Israeli and European scientists. The monoterpenoid numbering system can easily be adapted to describe open-ring cannabinoids such as cannabidiol (CBD) and derivatives as well as *iso*-THCs. The International Union of Pure and Applied Chemistry is yet to have a ruling on the system of numbers that should be used.

Many countries now have approved prescription drugs that contain phytocannabinoids that are synthetic products; among them is Marinol® (dronabinol), available in Germany and the United States. Cesamet™ (nabilone) which is not a phytcannabinoid but closely related is available in Canada, Mexico, the United Kingdom, and the United States as well [6].

As there are many challenges associated with extracting and purifying phytocannabinoids from cannabis due to close similarities in their physical and chemical properties, consequently, chemical synthesis of some cannabinoids became an attractive source for securing supply.

5.2 Synthesis of cannabigerolic acid

Cannabigerolic acid (CBGA) is one of the major cannabinoid acids found naturally within the cannabis plant in high concentrations, and it is the biosynthetic precursor of the major cannabinoids in cannabis. As described in Chapter 6, cannabidiolic acid (CBDA and cannabichromenic acid (CBCA) are formed by the oxidative cyclization of CBGA. On the other hand, tetrahydrocannabinolic acid (THCA) is biosynthesized by the isomerization of CBDA [7]. CBGA was detected around the same time as THCAA and CBDA in *C.sativa* in 1964 [8]. CBGA has shown different antiepileptic effects on seizures in mouse models of epilepsy and potentiated the anticonvulsant effects of clobazam [9].

The synthetic route for CBGA is one of the simplest reactions to date, which was mentioned in three separate reviews, over a span of 3 years. In this route, CBG is reacted with methyl magnesium carbonate (MMC), leading to a highly reactive intermediate, which immediately converts to CBGA. In this route, MMC acts as one of the most efficient carboxylating agents of the ortho position of phenolic compounds, facilitated by the phenolic group. In the case of CBG the ortho position of the other hydroxyl group present is not targeted by this agent due to steric hindrance [10–12]. This chemical reaction is shown in Figure 5.1.

5.3 Synthesis of CBG

CBG is produced in the plant material as its acidic form; CBG is then decarboxylated by heat to produce the "active" form [13]. CBG does not have much activity at the cannabinoid receptors [14, 15]. Studies indicated that CBG may have therapeutic potential in treating neurologic disorders (e.g., Huntington disease, Parkinson disease, and multiple sclerosis (MS)) and inflammatory bowel disease [16].

5.3.1 Synthesis of CBG from olivetol and geraniol using silica gel

Olivetol and geraniol are dissolved in dichloromethane and then added to a suspension of silica and BF_3-diethyl etherate in dichloromethane. This reaction mixture is stirred for 2 days and worked up. CBG is then purified using silica gel column chromatography and eluted with a mixture of ethyl acetate: petroleum ether (1: 9) in 29% yield [17]. This reaction can be seen in Figure 5.2.

5.3.2 Synthesis of CBG from olivetol and geraniol using anhydrous magnesium sulfate

Olivetol and BF_3-diethyl etherate are dissolved in chloroform at 20°C and then anhydrous magnesium sulfate is added to the mixture. Geraniol solution in chloroform is then added dropwise, and the reaction mixture was stirred for 3 h. After worked up, the crude product is subjected to flash column chromatography to produce CBG [18]. This reaction is shown in Figure 5.3.

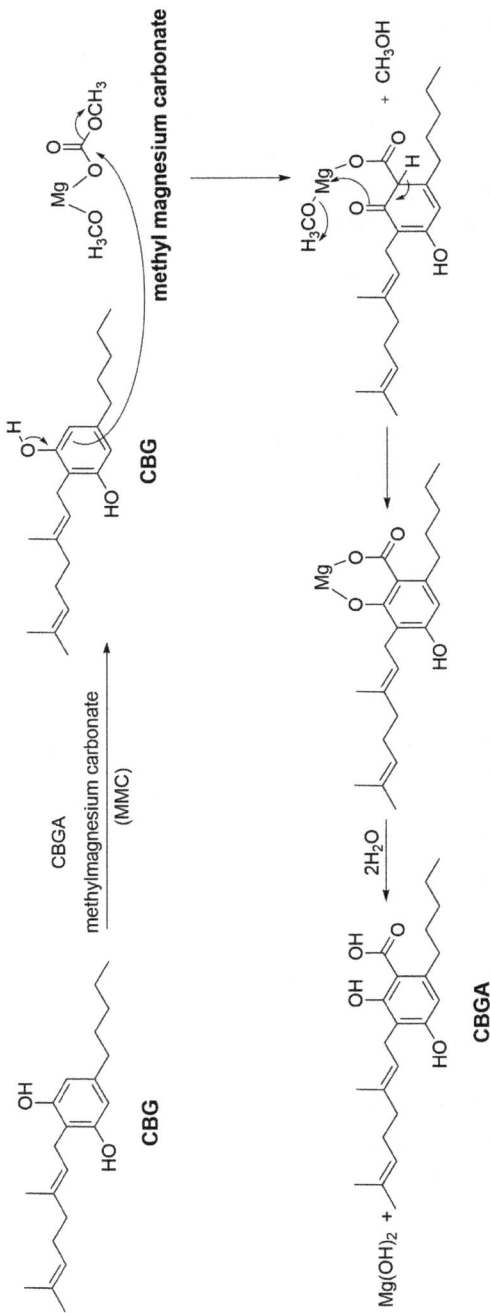

Figure 5.1: Synthesis of CBGA from CBG using methyl magnesium carbonate.

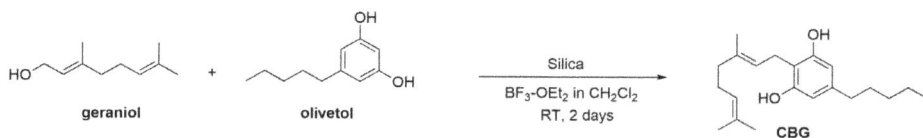

Figure 5.2: Synthesis of CBG using olivetol, geraniol, and silica gel.

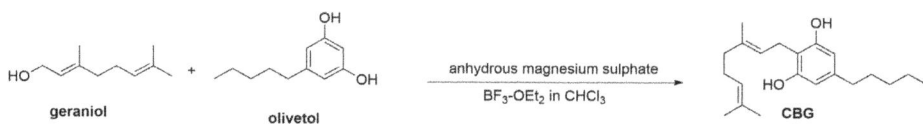

Figure 5.3: Synthesis of CBG using olivetol, geraniol, and anhydrous magnesium sulfate.

5.4 Synthesis of (−)-Δ^9-THC and CBD

Δ^9-THC, the main psychotropic ingredient isolated from *C. sativa* and its structure determined in 1964, represents the key drug target for the treatment of many diseases [19]. The biological activity of Δ^9-THC is attributed to its interaction with the endocannabinoid receptors CB1 and CB2 [20]. Δ^9-THC has many therapeutic applications and can be used as an anti-inflammatory in chronic neuropathic pain and the alleviation of spasticity associated with MS [21, 22]. In the United States, THC-based preparations have been approved by the Food and Drug Administration (FDA) under the trade names of nabilone (Cesamet®) and dronabinol (Marinol®) to be used for the treatment of nausea and vomiting side effects in patients receiving chemotherapy. In addition, dronabinol is prescribed to AIDS patients as an appetite stimulant.

CBD is considered the second most predominant cannabinoid in *C. sativa*. Contrary to Δ^9-THC, CBD exhibits no psychotropic effects. CBD is derived directly from the hemp plant and is an essential component of cannabinoids therapy.

In 1940, Adam et al. [23] isolated CBD, and in 1963 Mechoulam and Shvo [24,25] determined the structure, and in 1967, they determined the CBD structure in its entirety. CBD has been reported for the treatment of a wide variety of health issues with little direct interaction with the CB1 and CB2 receptors [14]. CBD has shown an effect in treating the harshest two forms of childhood epilepsy; the Dravet syndrome and the Lennox–Gastaut syndrome), especially in patients who do not respond to other antiepileptic medications [26]. Many studies proved that CBD was able to reduce the number of daily seizures or stop seizures altogether [27]. The FDA has approved a CBD-based liquid medication (Epidiolex®) as the first cannabis-derived medicine for the treatment of these conditions [28]. Nabiximol (Sativex®), an oral spray that is currently available in Canada and several other countries for treatment of MS spasticity, combines Δ^9-THC and CBD in a 1:1 ratio [6].

Due to the similarities in the synthetic procedures to produce both Δ^9-THC and CBD, their synthetic routes are discussed together. There is a total of four general synthetic methodologies that have been used to date to produce Δ^9-THC, namely decarboxylation of Δ^9-THCAA, semisynthetically, and the total synthesis pathway.

5.4.1 Synthesis of Δ^9-THC from its natural precursor (Δ^9-THC-acid A)

First, (–)-*trans*-Δ^9-THC does not occur naturally in the plant; its acid precursor, Δ^9-THCAA, is the primary form of (–)-*trans*-Δ^9-THC present in the plant. Δ^9-THCAA is synthesized naturally from CBGA and can be found in the glandular trichomes of the flowers and leaves (approximately 90% of (–)-*trans*-Δ^9-THC comes from the Δ^9-THCAA found in these parts of the plants) [29, 30]. While (–)-*trans*-Δ^9-THC has psychotic and euphoric effects on humans, Δ^9-THCAA has no such effects (1). Δ^9-THCAA can be decarboxylated by heat, as in baking, or smoking, yielding pure (–)-*trans*-Δ^9-THC [31]. The chemical reaction can be seen in Figure 5.4.

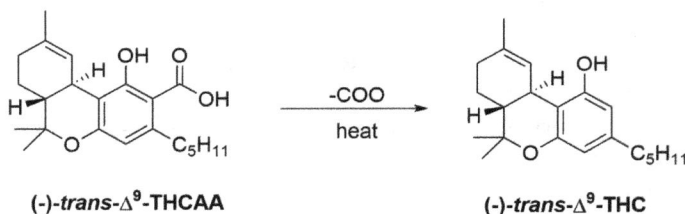

Figure 5.4: Decarboxylation of Δ^9-THCAA to Δ^9-THC.

5.4.2 Semisynthesis of Δ^9-THC

There is a total of two different semisynthesis pathways that can be used to produce Δ^9-THC. The first method uses CBD as the precursor to prepare Δ^9-THC.

5.4.2.1 Semisynthesis of Δ^9-THC from CBD using boron trifluoride diethyl etherate

In this method, CBD is subjected to boron trifluoride diethyl etherate (BF$_3$ · Et$_2$O) in DCM. This reaction resulted in the formation of two different products, namely Δ^9-THC (major product) and Δ^8-*iso*-THC (minor product), with yields of 60% and 13%, respectively [19, 31, 32]. This chemical reaction is shown in Figure 5.5.

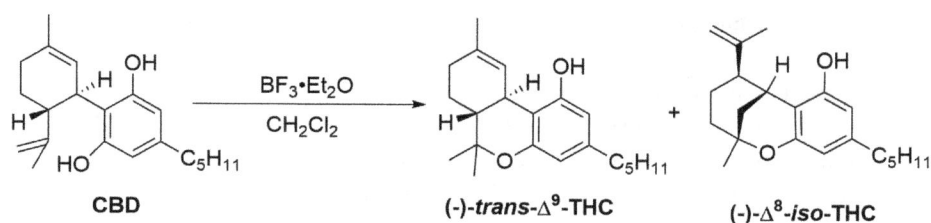

Figure 5.5: Synthesis of Δ^9-THC from CBD using BF$_3$·Et$_2$O.

5.4.2.2 Semisynthesis of Δ^9-THC from CBD using strong acidic medium

The second semisynthesis pathway also utilizes CBD as a precursor to prepare (–)-*trans*-Δ^9-THC. The first step of the reaction effectively converts CBD into Δ^8-THC through the use of strong acidic conditions. The acidic conditions induce the sterically hindered hydroxyl group in CBD to cyclize into an ether group, forming Δ^8-THC. The next step converts Δ^8-THC into Δ^9-THC through the use of chlorination followed by dehydrochlorination. Reacting Δ^8-THC with HCl in zinc chloride allows for the chlorination of Δ^8-THC at the double bond to form the 9-chloro-hexahydrocannabinol. Next, upon treatment of the intermediate with potassium *tert*-amylate, the double bond forms in the least sterically hindered position; this results in the formation of Δ^9-THC [19, 33–37]. The conversion of CBD into (–)-*trans*-Δ^9-THC is shown in Figure 5.6.

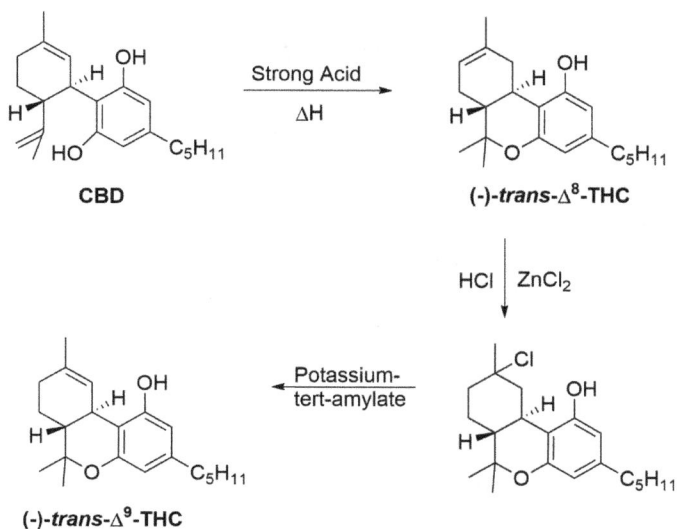

Figure 5.6: Synthesis of Δ^9-THC from CBD using potassium *tert*-amylate and strong acidic medium.

5.4.2.3 Semisynthesis of Δ⁹-THC from CBD using trifluoroacetic acid

CBD is dissolved in dry toluene at 0 °C. To this solution, trifluoroacetic acid (TFA) is added and the mixture was stirred for 6 h. Temperature was warmed up to 20 °C and reaction mixture was worked up. After evaporation of the solvent, the crude mixture is subjected to silica gel flash column chromatography to produce Δ⁹-THC [18]. This reaction is shown in Figure 5.7.

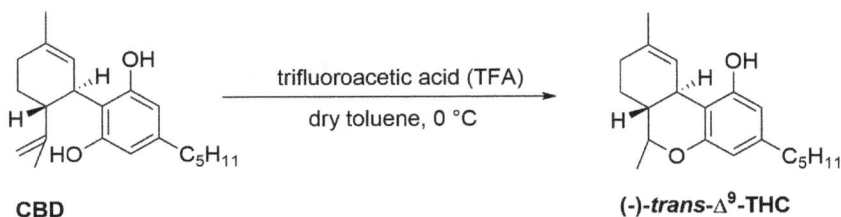

CBD   **(-)-*trans*-Δ⁹-THC**

Figure 5.7: Scheme for the semisynthesis of Δ⁹-THC from CBD.

5.4.3 Total synthesis of Δ⁹-THC/CBD

There is a total of three different synthetic routes to produce Δ⁹-THC and CBD.

5.4.3.1 Total synthesis of Δ⁹-THC from olivetol and both the *trans* and *cis* isomers of *p*-mentha-2,8-dien-1-ol

The first method, reported by Petrzilka et al., involves the reaction between olivetol and both the *trans* and *cis* isomers of *p*-mentha-2,8-dien-1-ol. Using a strongly acidic condition, this reaction forms Δ⁸-THC. Then, using the same method described above for the second semisynthesis method, chlorination/dehydrochlorination is utilized again in order to move the double bond from the Δ⁸ position to the Δ⁹ position [38, 39]. The total synthesis of Δ⁹-THC can be seen in Figure 5.8.

5.4.3.2 Dethe synthesis of Δ⁹-THC and CBD

The next total synthesis route is termed the Dethe synthesis. This synthetic route can be used to produce both CBD and Δ⁹-THC. For both cannabinoids, olivetol is reacted with 3-methyl-6-(prop-1-en-2-yl)cyclohex-2-en-1-ol along with boron trifluoride diethyl etherate (BF₃-OEt₂). However, the concentration of BF₃-OEt₂ determines which compound is generated. In order to generate CBD, 10 mol% BF₃-OEt₂ was used, while to

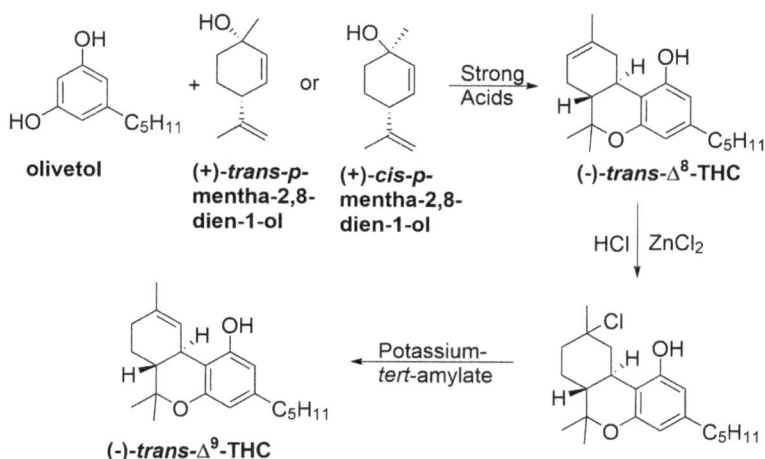

Figure 5.8: Synthesis of Δ⁹-THC from olivetol and both the *trans* and *cis* isomers of *p*-mentha-2,8-dien-1-ol using a strongly acidic condition.

generate Δ⁹-THC, 50 mol% of $BF_3\text{-}OEt_2$ was used [40]. The Dethe synthesis of Δ⁹-THC can be seen in Figure 5.9.

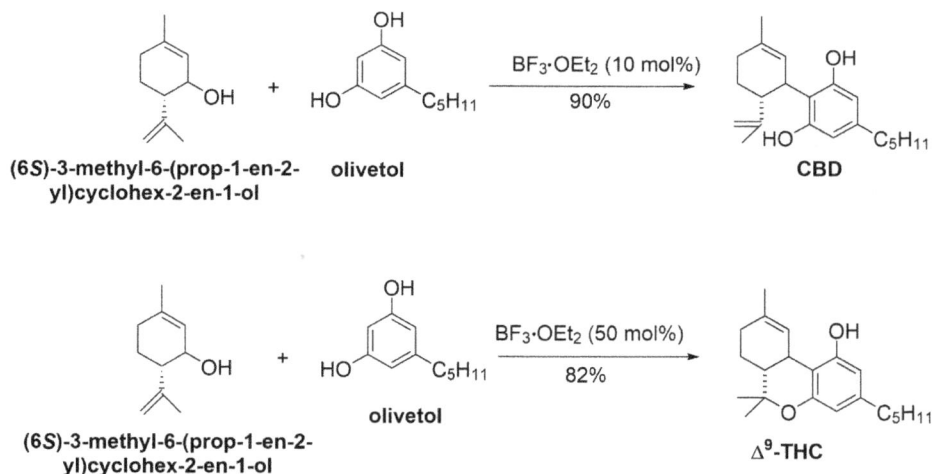

Figure 5.9: The Dethe synthesis of CBD and Δ⁹-THC.

5.4.3.3 Stoss and Merrath synthesis of Δ⁹-THC

In order to increase the % yield of (–)-*trans*-Δ⁹-THC, Stoss and Merrath first optimized the synthesis of (–)-*trans*-6-hydroxy-CBD which was then used to synthesize (–)-*trans*

-Δ⁹-THC. This intermediate is a solid and is purified through crystallization. Although it is a two-step synthesis, the overall yield of (–)-*trans*-Δ⁹-THC is better because of no abnormal derivatives as well as any *bis*-adduct were produced during the reaction and therefore better clean-up procedure for higher overall yield [41]. This chemical reaction is shown in Figure 5.10.

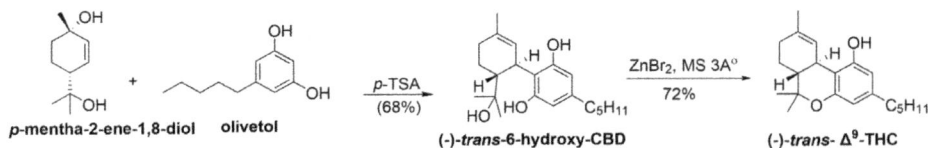

Figure 5.10: Δ⁹-THC synthesis scheme proposed by Stoss and Merrath.

5.4.3.4 Synthesis of Δ⁹-THC using (–)-verbenone

(–)-verbenone was converted to *cis*-chrysanthenol which was reacted with olivetol to produce (–)-*trans*-Δ⁹-THC [42–44]. The reaction can be seen in Figure 5.11.

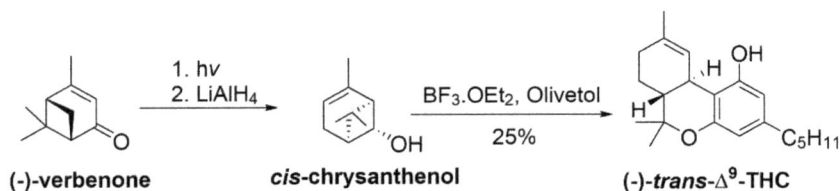

Figure 5.11: Synthesis of (–)-*trans*-Δ⁹-THC using (–)-verbenone.

The Diels–Alder reaction of diene and 3-acryloyloxazolidin-2-one in the presence of catalyst produced cycloadduct (1S,6S)-3-methyl-6-(2-oxooxazolidine-3-carbonyl) cyclohex-2-en-1-yl acetate which was selectively cleaved with LiOBn to give benzyl (2S)-2-acetoxy-4-methylcyclohex-3-ene-1-carboxylate. To this when methyl magnesium bromide was added at 0°C, (1S,6S)-6-(2-hydroxypropan-2-yl)-3-methylcyclohex-2-en-1-ol was produced. When olivetol was added to this intermediate along with *p*-toluene sulphonic acid in dichloromethane, (1′S,2′S)-2′-(2-hydroxypropan-2-yl)-5′-methyl-4-pentyl-1′,2′,3′,4′-tetrahydro-[1,1′-biphenyl]-2,6-diol was obtained, which was cyclized in the presence of zinc bromide and magnesium sulfate to give (+)-*trans*-Δ⁹-THC. Spectroscopic data of (+)-*trans*-Δ⁹-THC was similar to the published data of natural (–)-*trans*-Δ⁹-THC except opposite in optical rotation [45]. This chemical reaction is shown in Figure 5.12.

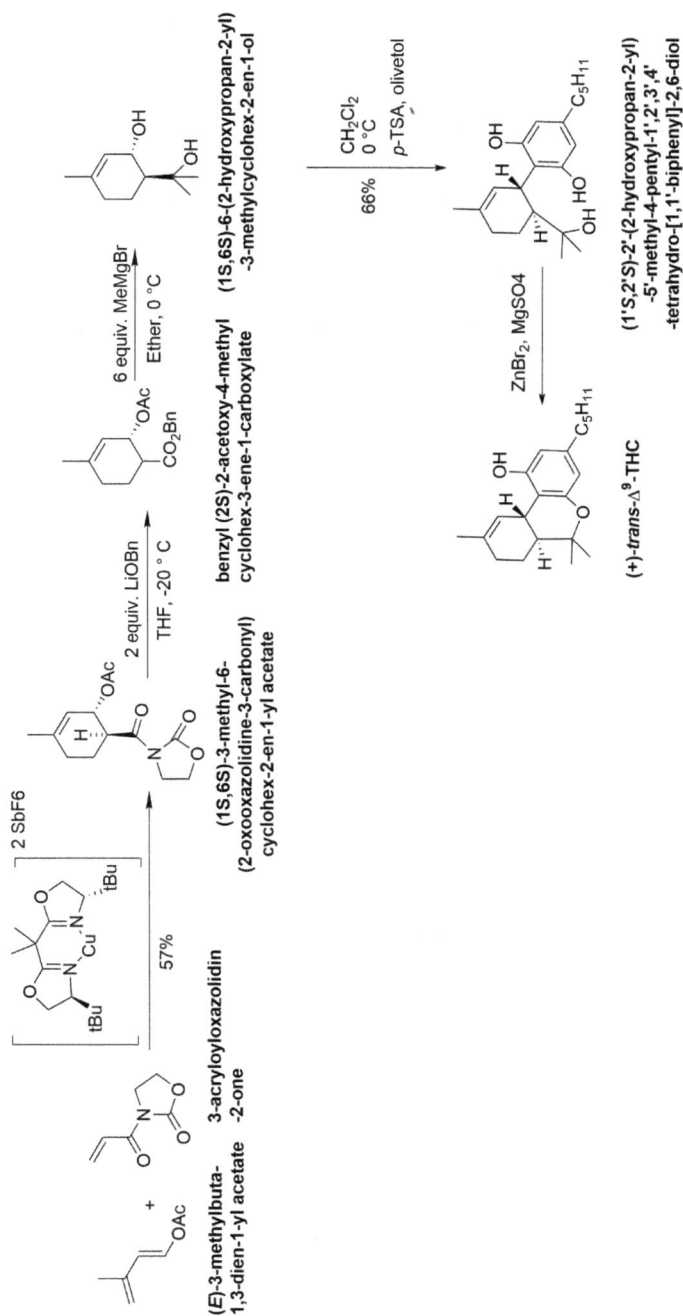

Figure 5.12: The Diels–Alder reaction for the synthesis of (+)-*trans*-Δ⁹-THC.

122 —— Waseem Gul et al.

5.4.3.5 Semisynthesis of Δ⁹-tetrahydrocannabivarin from cannabidivarin

Δ⁹-Tetrahydrocannabivarin (Δ⁹-THCV) is a propyl homologue of Δ⁹-THC and has recently been reported to have diverse pharmacology and showed tissue-dependent effects [14]. Recent studies have shown that Δ⁹-THCV exhibits an antagonist action at CB1 and CB2 receptors in whole mouse brain membranes and recombinant cells, respectively [46]. Δ⁹-THCV also showed Δ⁹-THC-like properties as catalepsy in a mouse ring test, even at a lower potency [47].

Cannabidivarin (CBDV) was dissolved in dry toluene at 0 °C. To this solution, TFA was added and the mixture was stirred for 6 h. Temperature was then warmed up to 20 °C and the reaction mixture was worked up. After evaporation of the solvent, the crude mixture was subjected to silica gel flash column chromatography to produce Δ⁹-THCV [18]. This reaction can be seen in Figure 5.13.

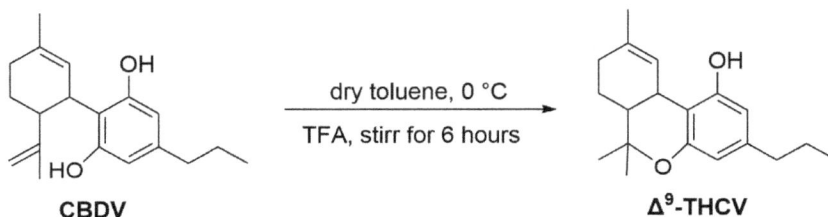

Figure 5.13: Scheme for semisynthesis of Δ⁹-THCV from CBDV.

5.4.3.6 Total synthesis of CBD from olivetol

To a solution of olivetol in chloroform (at 20 °C), anhydrous magnesium sulfate and BF₃-diethyl etherate were added. To this mixture, p-mentha-1,8-diene-ol dissolved in chloroform was added dropwise. After 3 h, the reaction mixture was worked up, and the crude product was subjected to silica gel flash column chromatography to produce CBD as white solid [18]. This chemical reaction is shown in Figure 5.14.

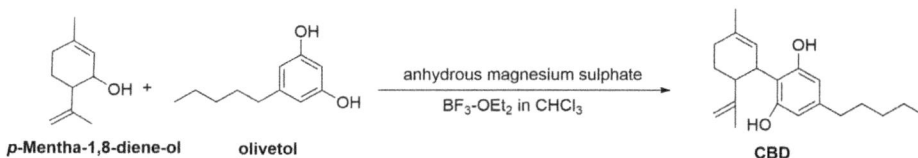

Figure 5.14: Synthesis of CBD from olivetol.

5.4.3.7 Total synthesis of CBDV from divarinol

CBDV, the propyl homologue of CBD, is a nonpsychotropic cannabinoid. Many studies have reported CBDV as an emerging candidate with the potential to reduce seizures [48–50].

Divarinol was dissolved in chloroform at 20 °C, and then anhydrous magnesium sulfate and BF$_3$-diethyl etherate were added. To this mixture, p-mentha-1,8-diene-ol dissolved in chloroform was then added dropwise. After 3 h, the reaction mixture was worked up, and the crude product was subjected to silica gel flash column chromatography to produce CBDV as white solid [18]. This chemical reaction is shown in Figure 5.15.

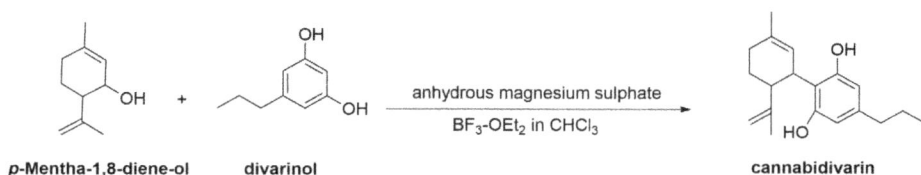

Figure 5.15: Synthesis of CBDV from divarinol.

5.5 Synthesis of Δ^9-THCAA

Δ^9-THCAA is the precursor of Δ^9-THC, the psychoactive component of cannabis [51, 52]. Δ^9-THCAA is biosynthesized from CBGA by the enzyme THCA synthase which mediates the conversion of CBGA to THCAA [51–53]. Δ^9-THC is reacted with MMC, leading to a highly reactive intermediate which immediately reverts to Δ^9-THCAA. In this route, MMC acts as one of the most efficient carboxylating agents of phenolic compounds [11, 12, 54]. This chemical reaction is shown in Figure 5.16.

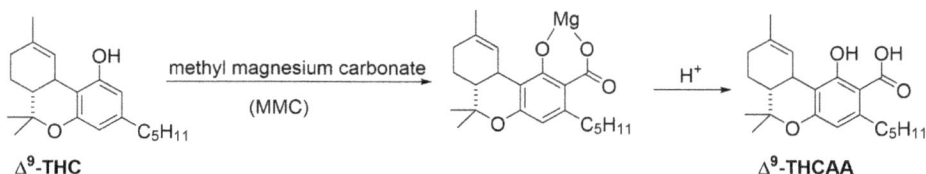

Figure 5.16: Synthesis of Δ^9-THCAA from Δ^9-THC using methyl magnesium carbonate.

5.6 Synthesis of CBDA

CBDA was the first cannabinoid acid discovered and isolated in 1955 by Krejci and Šantavý [55]. CBDA is one of the major cannabinoid acids found naturally within the cannabis plant, with the highest concentrations in the hemp plants (CBD chemovar). CBDA does not exert psychotropic effects but possesses a variety of pharmacological effects including antimicrobial activity [56]. CBDA is formed biosynthetically from CBGA via CBDA synthase enzyme [1].

The synthetic route for this cannabinoid acid starting from CBD is the same process as previously shown with CBGA and Δ^9-THCAA, where CBD is reacted with MMC, through a highly reactive intermediate, which immediately reverts to CBDA. This reaction is facilitated by only one of the phenolic groups [10–12]. This chemical reaction is shown in Figure 5.17.

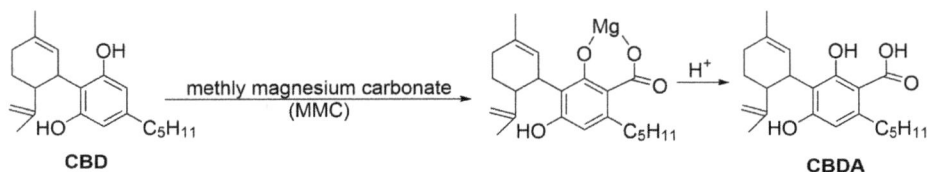

Figure 5.17: Conversion of CBD into CBDA using methyl magnesium carbonate.

5.7 Synthesis of cannabinol

Cannabinol (CBN) is a unique minor cannabinoid found in cannabis and structurally related to Δ^9-THC. Due to air oxidation, Δ^9-THC undergoes aromatization of the terpene ring, possibly through several hydroxy intermediates which eventually produce CBN [57]. It was isolated from hashish at the end of the nineteenth century and structurally elucidated in 1940 [58]. CBN is partially CB1 and CB2 agonist [59].

CBN has shown anxiolytic and appetite stimulant effects but it is less active than CBD and Δ^9-THC, respectively [16]. Finally, CBN has been reported to have anticancer activity [60].

5.7.1 Synthesis of CBN using CBC as a starting material

In 2019, Caprioglio et al. [61] dissolved CBC in toluene and while stirring, added iodine. This mixture was refluxed for 3 h, and the progress of the reaction was monitored by thin layer chromatography using (petroleum ether: ethyl acetate; 9:1), which indicated the completion of the reaction. After workup of the reaction mixture and solvent evaporation, the crude reaction mixture was purified by silica gel gravity column chromatography using petroleum ether: ethyl acetate (95:5) to produce CBN as brown oil. This chemical reaction is shown in Figure 5.18.

5.7.2 One-pot total synthesis of CBN

Olivetol was dissolved in toluene, and citral and *n*-butylamine were added. After refluxing the mixture overnight, it was cooled to room temperature. Dowex 50 W × 8 (200 mg) was then added, and the mixture was stirred for 10 min at room temperature, followed by filtration over celite. Iodine was added to this filtered solution and refluxed for 3 h. After workup, the residue was purified by silica gel gravity column using petroleum ether: ethyl acetate (95:5) to afford CBN as brown oil [61]. This chemical reaction is shown in Figure 5.19.

5.7.3 Proposed pathway of the oxidation of Δ⁹-THC to CBN

Turner and ElSohly [62] proposed a possible decomposition pathway of Δ^9-THC and its Δ^8-isomer to CBN through epoxy and hydroxylated intermediates. This pathway is shown in Figure 5.20.

5.7.4 Synthesis of CBN from CBD

CBD was dissolved in dry toluene and iodine was added to it. The solution was refluxed for 18 h. After work up, the residue was purified using silica gel column chromatography to produce CBN [18]. This chemical reaction is shown in Figure 5.21.

5.8 Synthesis of cannabielsoin

In 1973, cannabielsoin (CBE) was identified by Bercht et al. [63]. CBE could be generated through the cyclization of CBD which occurs between electron-rich alkenyl or hydroxyl structures [64].

In 1974, CBE was fully synthesized from CBD, through a relatively simple, three-step synthetic route. CBD was acetylated through the use of pyridine and acetic anhydride at room temperature, forming CBD diacetate. This intermediate was then subjected to *m*-chloroperbenzoic acid in chloroform, yielding a mixture of three different epoxides. Two of the epoxides formed were monoepoxides, while the third epoxide was a diepoxide. The two monoepoxides were mixed and subjected to 2% sodium

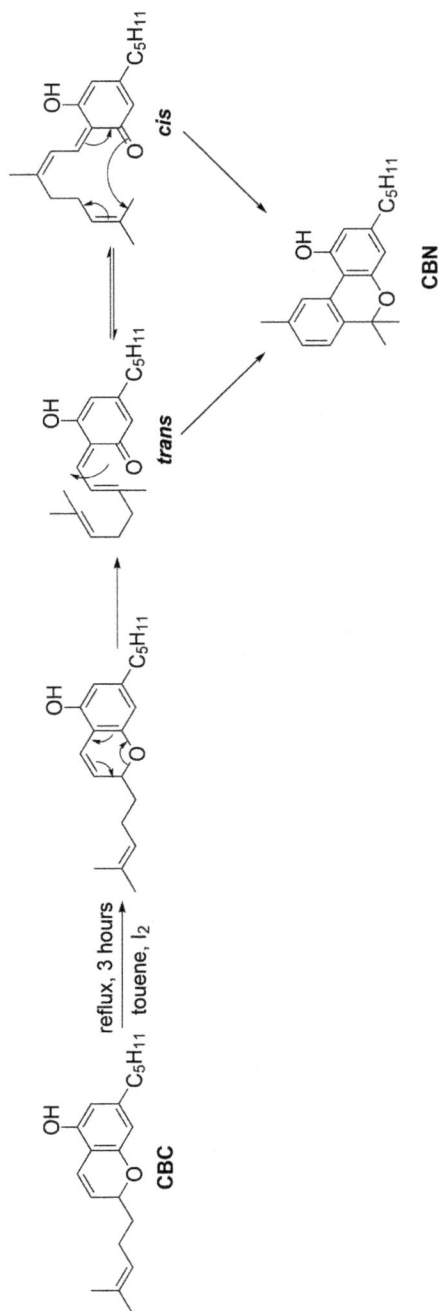

Figure 5.18: Scheme for the synthesis of CBN from CBC.

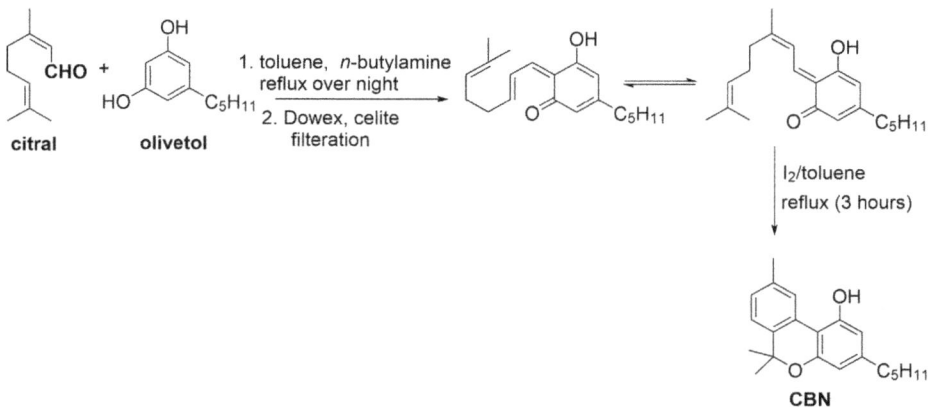

Figure 5.19: Scheme for the synthesis of CBN using citral and olivetol.

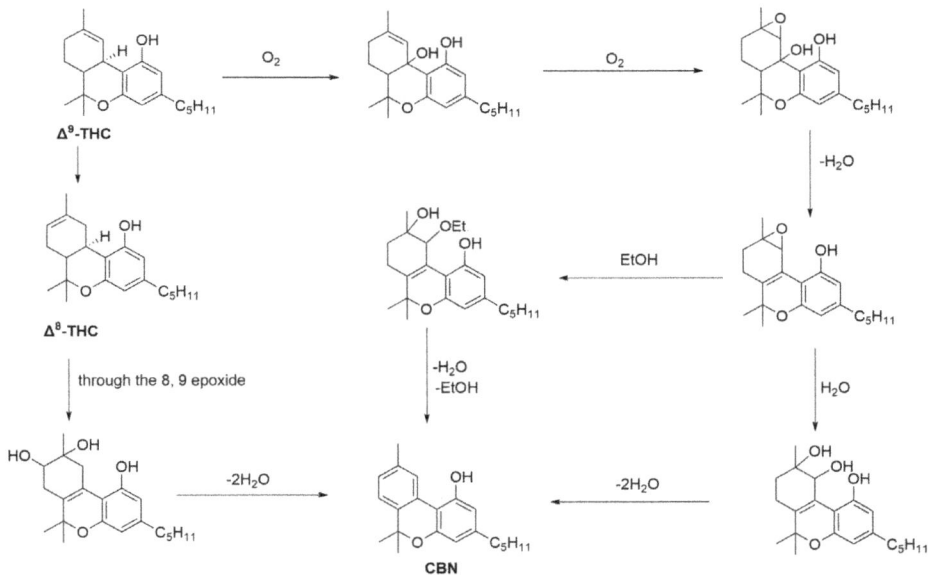

Figure 5.20: A possible decomposition pathway of Δ^9-THC to CBN.

hydroxide in methanol:water (1:1) at room temperature. Purification by silica gel preparative TLC using ether: petroleum ether, 1:1 solvent system produced pure CBE, and a derivative of Δ^9-THC in an approximate 75% yield in a ratio of 4:1 [65]. This chemical reaction is shown in Figure 5.22.

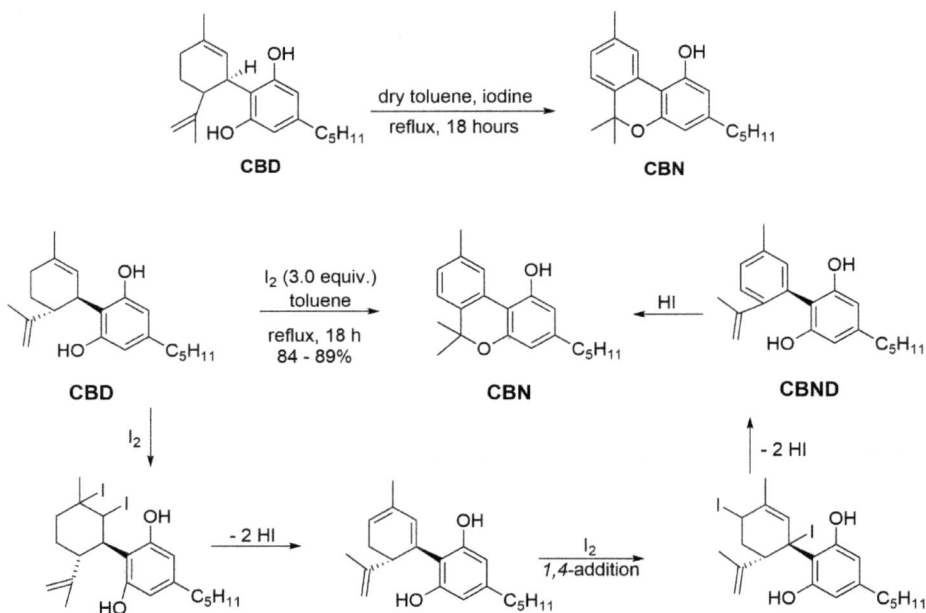

Figure 5.21: Synthesis of CBN from CBD.

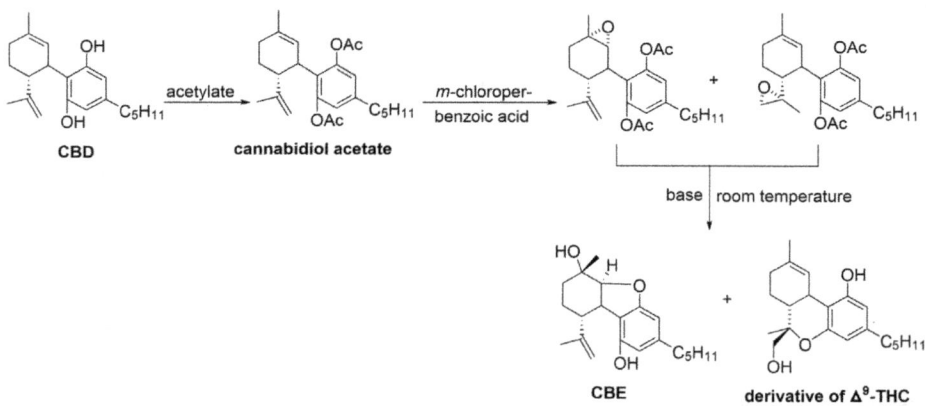

Figure 5.22: Synthesis of CBE from CBD.

5.9 Synthesis of cannabichromene

Cannabichromene (CBC) was first isolated from *C. sativa* in 1966 [66, 67]. Many studies have reported the use of CBC in the treatment of epilepsy and as anti-inflammatory therapeutic agent [68, 69].

5.9.1 Synthesis of CBC

CBC, a non-narcotic major constituent of cannabis leaves, was discovered and reported by Claussen et al. [70] and Gaoni and Mechoulam in 1966 [66]. CBC is one of the major cannabinoids in cannabis and is the second most abundant cannabinoid in some strains of marijuana growing in the United States [71]. CBC has been reported to potentiate the analgesic effect of Δ^9-THC in the mouse tail-flick assay [72]. In 1966, CBC was scientifically shown to induce sedation and ataxia in canines [66]. In addition, it has also been shown to have anti-inflammatory properties *in vivo* and *in vitro* [69, 73, 74]. There is a multitude of synthetic routes for the synthesis of CBC.

5.9.2 Synthesis of CBC from CBG by Mechoulam et al

In 1968, Mechoulam and Gaoni reported the first total synthesis of CBC. In their reaction, CBG was subjected to dehydrogenation through the use of chloranil in benzene. The products of this reaction were CBC (45% yield) and a tetracyclic diether (15% yield). This product was optically identical to the CBC isolated form of cannabis, having identical stereospecific and spectroscopic characteristics (determined through vapor phase chromatography, IR, NMR, and MS spectra) and the same R_f on thin-layer chromatography. The melting point for the compound ranged from 106 °F to 107 °F [75]. This chemical reaction is shown in Figure 5.23.

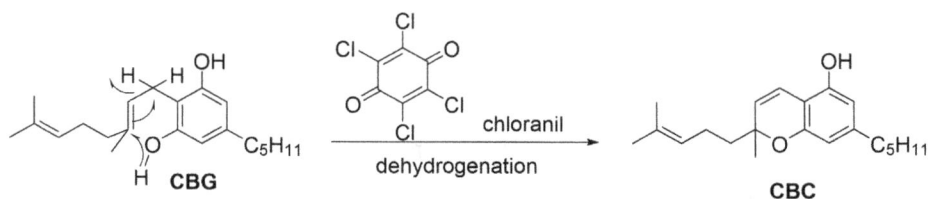

Figure 5.23: The first total synthesis route for CBC from CBG proposed by Mechoulam and Gaoni.

5.9.3 Synthesis of CBC from olivetol and citral as starting materials using *t*-butylamine

ELSohly et al. dissolved olivetol and citral in toluene and *t*-butylamine and refluxed the reaction mixture at 110 °C for 7 h. Sodium borohydride reduction was used for the conversion of unreacted citral into its corresponding alcohol as the separation of unreacted citral from CBC was difficult because of the close R_f on silica gel. Column chromatography on 1% NaOH impregnated silica gel yielded CBC in 62% yield [76]. This chemical reaction is shown in Figure 5.24.

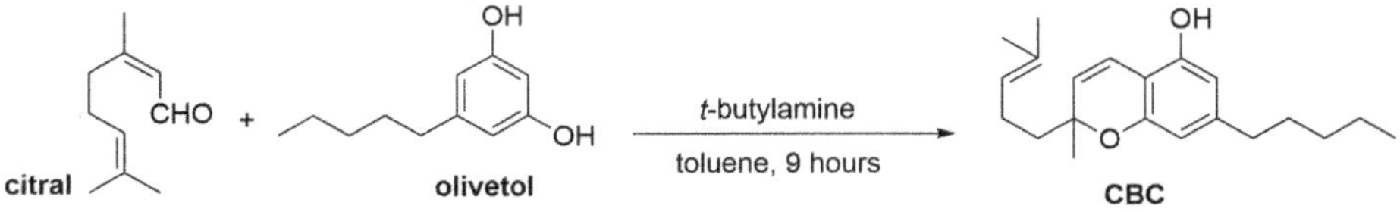

Figure 5.24: The first total synthesis route for CBC from CBG proposed by ElSohly et al.

5.9.4 Synthesis scheme of CBC from CBG proposed by Cardillo, Cricchio, and Merlini

In 1968, Cardillo, Cricchio, and Merlini utilized 2,3-dichloro-5,6-dicyanobenzoquinone (DDQ) to synthesize a variety of natural chromenes. This synthesis pathway was conducted through the cyclodehydrogenation of the appropriate isoprenylphenols with DDQ to yield the corresponding products [77]. This report was based on the hypothesis of Ollis and Sutherland [78] in 1961 that DDQ abstracts a hydride ion; this leads to the production of a quinone methide. As this intermediate is highly unstable, it immediately converts to chromene. Based on this deduction, Campbell et al. [79], in 1966, conducted the first reproduction of this reaction in vitro through the modification of the hypothesis of Ollis and Sutherlands (3). In this reaction, CBG was subjected to cyclodehydrogenation with DDQ, which resulted in the formation of the quinone methide; this was immediately converted into CBC [79]. This chemical reaction is shown in Figure 5.25.

5.9.5 Synthesis of CBC from olivetol and citral as starting materials using ethylenediamine diacetate

Olivetol was dissolved in toluene and citral and ethylenediamine diacetate was added to it. The reaction mixture was stirred for 6 h at 130 °C. The mixture was then cooled down to room temperature and the solvent evaporated. The residue was subjected to silica gel flash column chromatography to produce CBC as brown, yellow oil [18]. This chemical reaction is shown in Figure 5.26.

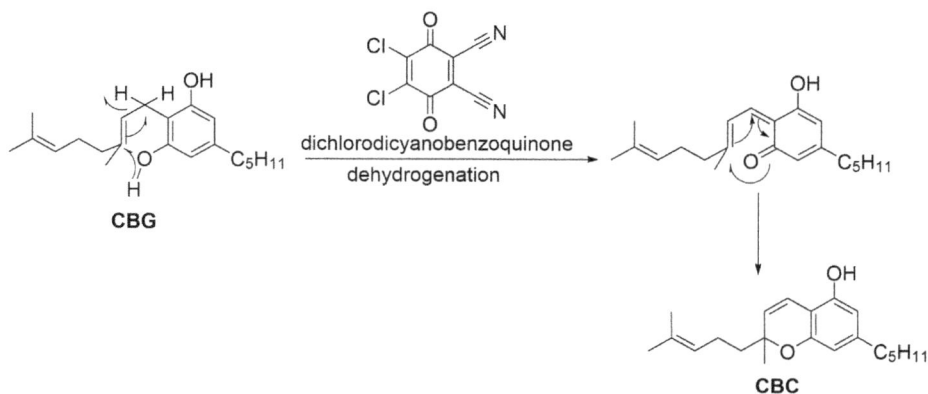

Figure 5.25: Synthesis scheme of CBC from CBG is proposed by Cardillo, Cricchio, and Merlini.

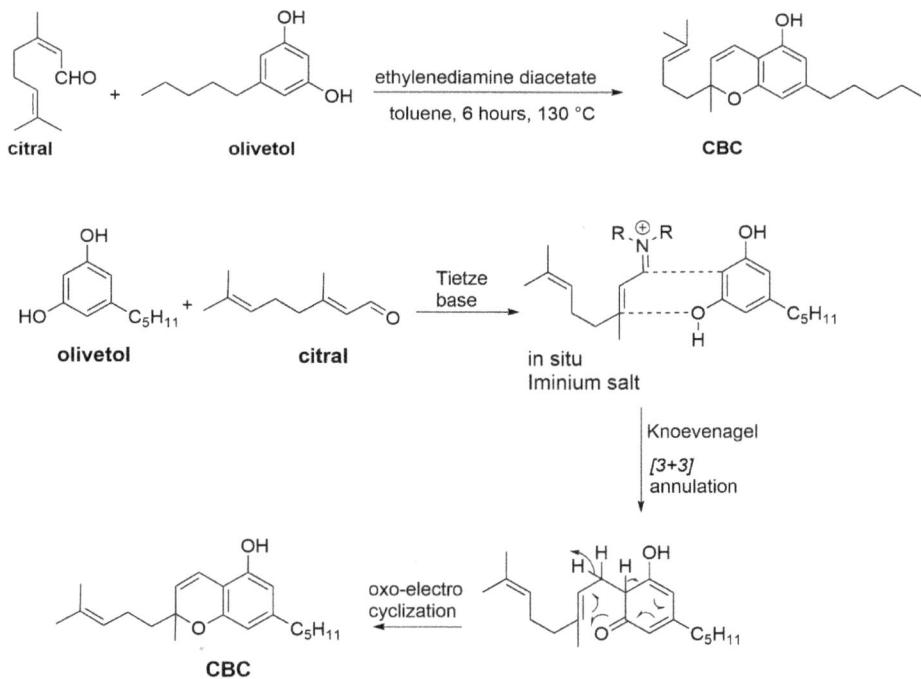

Figure 5.26: Synthesis scheme of CBC from citral and olivetol.

5.10 Proposed synthesis of cannabicyclol

Cannabicyclol (CBL) was isolated from hashish by Krote and Sieper in 1964 [80], and its structure was determined by Mechoulam and Gaoni in 1967 [81]. In 1970, Mechoulam proposed the natural formation of CBL from CBC through cyclization [1].

5.10.1 Synthesis of CBL from pyrolysis of CBC in the presence of silica gel

In 2019, a proposed mechanism for the conversion of CBC to CBL was theorized, involving a concerted process. The acidic tautomer of CBC, consisting of the resorcinol moiety, was generated through the thermolysis of the cannabinoid. Next, it was subjected to acidic electrophilic activation leading to the production of a cyclopropane intermediate. The reaction was later completed through an electrophilic Markovnikov addition to the homoisoprenyl terminal double bond. The termination of the reaction is due to the regeneration of the aromaticity of the compound through both decomplexation and tautomerization. Albeit the theoretical nature of this reaction, this is highly plausible, due to the generation of a tertiary cation, that is later converted into an oxonium ion through the opening of the cyclopropane ring [61].

Caprioglio et al. [61] adsorbed CBC on to silica and microwave was used to heat it to 150 °C for 210 min. TLC showed no starting material was left (petroleum ether: ethyl acetate, 9:1). The crude material was subjected to column chromatography to produce brown oil which was purified by reversed-phase HPLC producing CBL in 31.2% yield. Upon spectral analysis, this compound was identical to the data reported in literature. This chemical reaction is shown in Figure 5.27.

5.10.2 Synthesis of CBL from CBC

CBC was dissolved in dichloromethane, and TFA was added dropwise and while the temperature was maintained at 0 °C. The reaction mixture was stirred for 18 h, and then the temperature warmed up to 20 °C. The reaction mixture was worked up and the crude mixture was purified with silica gel flash column chromatography to produce CBL as white solid [18]. This chemical reaction is shown in Figure 5.28.

5.11 Synthesis of CBT from CBC

CBT is an important cannabinoid produced by *C. sativa* and is often found at surprisingly high levels in CBD rich extracts. CBT was isolated from Lebanese hashish

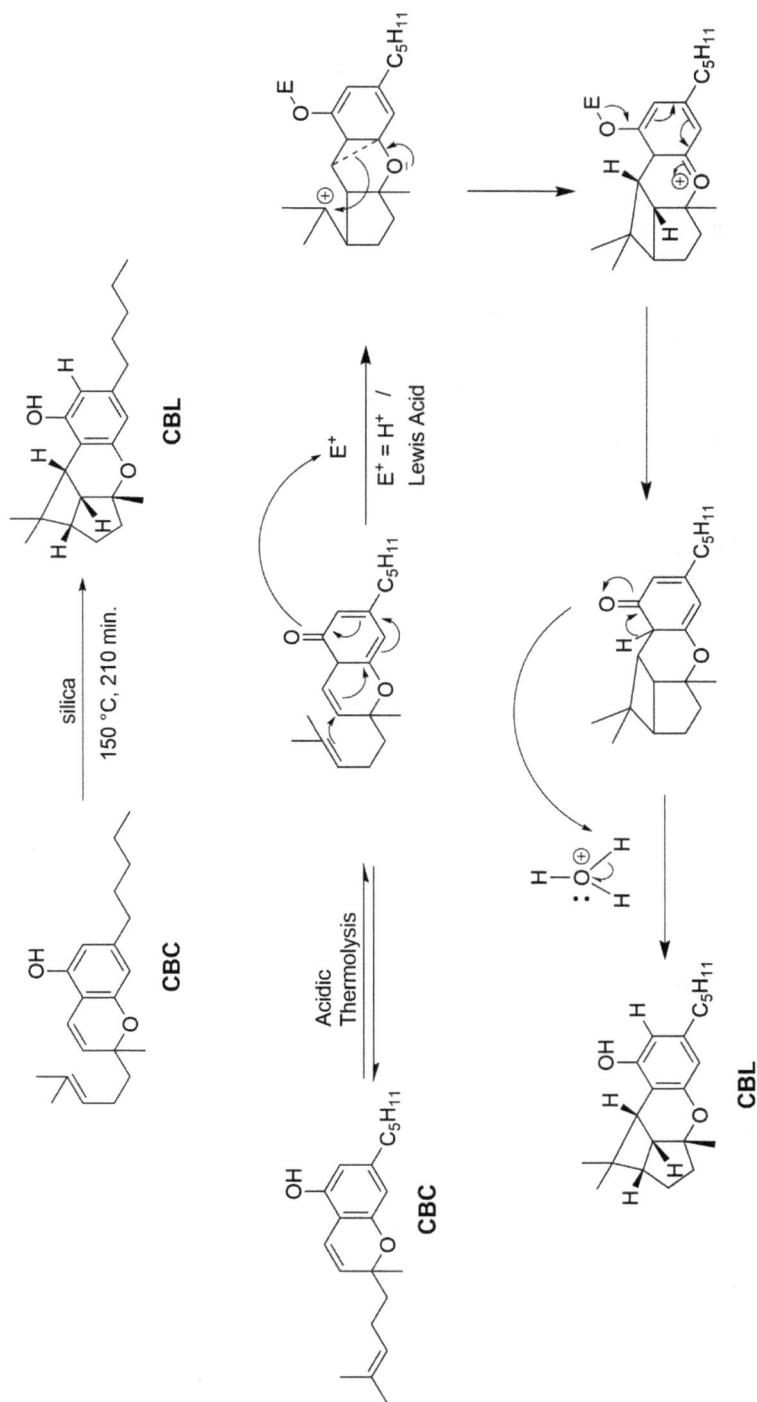

Figure 5.27: Synthesis of CBL from CBC using acidic thermolysis.

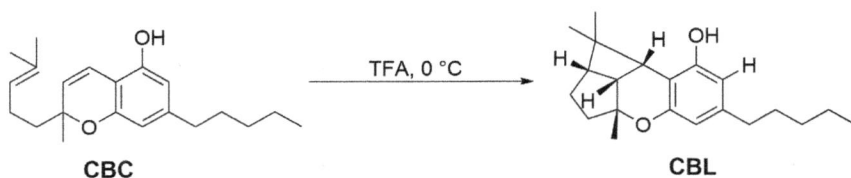

Figure 5.28: Synthesis of CBL starting from CBC using TFA.

by Berch et al [82]. It is not psychoactive, but was found to reduce intraocular pressure when tested on rabbits [83].

5.11.1 Synthesis of CBT using CBC as starting material (acidic thermolysis)

Caprioglio et al. [61] adsorbed CBC onto silica and microwave was used to heat it to 150 °C for 210 min. TLC confirmed the complete consumption of CBC (petroleum ether: ethyl acetate, 9:1). The crude material was subjected to column chromatography to produce brown oil which on further purification by reversed-phase HPLC produced CBT in 17.8% yield. Upon spectral analysis, synthesized CBT data were identical to the data reported in literature [61]. This chemical reaction is shown in Figure 5.29.

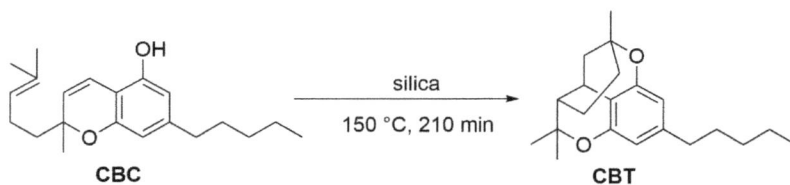

Figure 5.29: Synthesis of CBT using CBC as starting material (acidic thermolysis).

5.11.2 Synthesis of CBT using CBC as starting material (TFA as acid)

CBC was dissolved in dichloromethane, TFA was added dropwise at a temperature maintained at 0 °C, and the reaction mixture stirred for 18 h. The temperature was then warmed up to 20 °C, the reaction mixture was worked up, and the crude mixture was purified with silica gel flash column chromatography to produce CBT as white solid [18]. This chemical reaction is shown in Figure 5.30.

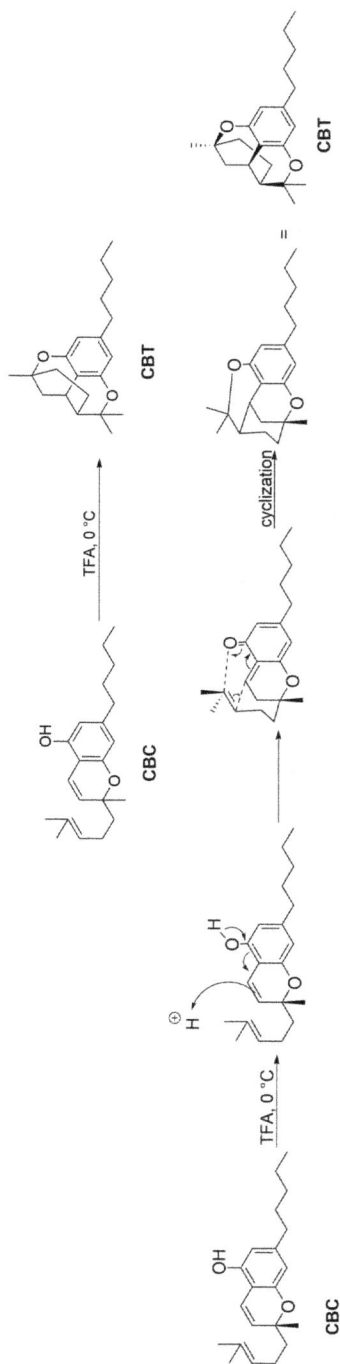

Figure 5.30: Synthesis of CBT from CBC.

References

[1] Mechoulam R. Marihuana chemistry: Recent advances in cannabinoid chemistry open the area to more sophisticated biological research. Science 1970; 168: 1159–1166.

[2] Radwan MM, Chandra S, Gul S, ElSohly MA. Cannabinoids, phenolics, terpenes and alkaloids of cannabis. Molecules 2021; 26: 2774.

[3] Mechoulam R, Gaoni Y. Fortschritte der chemie organischer naturstoffe/progress in the chemistry of organic natural products/progrès dans la chimie des substances organiques naturelles. Vienna, Springer, 1967.

[4] Hepler RS, Frank IR. Marihuana smoking and intraocular pressure. Jama 1971; 217: 1392–1392.

[5] Sallan SE, Zinberg NE, Frei III E. Antiemetic effect of delta-9-tetrahydrocannabinol in patients receiving cancer chemotherapy. New England Journal of Medicine 1975; 293: 795–797.

[6] Di Marzo V. New approaches and challenges to targeting the endocannabinoid system. Nature Reviews Drug Discovery 2018; 17: 623–639.

[7] Fellermeier M, Zenk MH. Prenylation of olivetolate by a hemp transferase yields cannabigerolic acid, the precursor of tetrahydrocannabinol. FEBS Letters 1998; 427: 283–285.

[8] Gaoni Y, Mechoulam R. Structure+ synthesis of cannabigerol new hashish constituent. Royal Soc Chemistry Thomas Graham House, Science Park, Milton Rd, Cambridge . . ., 1964; 82-&.

[9] Anderson LL, Heblinski M, Absalom NL, Hawkins NA, Bowen MT, Benson MJ, Zhang F, Bahceci D, Doohan PT, Chebib M. Cannabigerolic acid, a major biosynthetic precursor molecule in cannabis, exhibits divergent effects on seizures in mouse models of epilepsy. British Journal of Pharmacology 2021; 178: 4826–4841.

[10] Razdan R. Recent advances in the chemistry of cannabinoids. Progress in organic chemistry. In: Carruthers W, Sutherland JK (Eds). London, Butterworths, 1973; 8: 78–101.

[11] Mechoulam R, Burstein SH. Marijuana: Chemistry, pharmacology, metabolism and clinical effects. Contributors-SH Burstein [and Others]. Academic Press, New York, 1973.

[12] Mechoulam R, McCallum NK, Burstein S. Recent advances in the chemistry and biochemistry of cannabis. Chemical Reviews 1976; 76: 75–112.

[13] Fellermeier M, Eisenreich W, Bacher A, Zenk MH. Biosynthesis of cannabinoids: Incorporation experiments with 13C-labeled glucoses. European Journal of Biochemistry 2001; 268: 1596–1604.

[14] Pertwee R. The diverse CB1 and CB2 receptor pharmacology of three plant cannabinoids: Δ9-tetrahydrocannabinol, cannabidiol and Δ9-tetrahydrocannabivarin. British Journal of Pharmacology 2008; 153: 199–215.

[15] Navarro G, Varani K, Reyes-Resina I, Sanchez de Medina V, Rivas-Santisteban R, Sanchez-Carnerero Callado C, Vincenzi F, Casano S, Ferreiro-Vera C, Canela EI. Cannabigerol action at cannabinoid CB1 and CB2 receptors and at CB1–CB2 heteroreceptor complexes. Frontiers in Pharmacology 2018; 9: 632.

[16] Nachnani R, Raup-Konsavage WM, Vrana KE. The pharmacological case for cannabigerol. Journal of Pharmacology and Experimental Therapeutics 2021; 376: 204–212.

[17] Baek S-H, Du Han S, Yook CN, Kim YC, Kwak JS. Synthesis and antitumor activity of cannabigerol. Archives of Pharmacal Research 1996; 19: 228–230.

[18] Nguyen G-N, Jordan EN, Kayser O. Synthetic strategies for rare cannabinoids derived from *Cannabis sativa*. Journal of Natural Products 2022; 85: 1555–1568.

[19] Gaoni Y, Mechoulam R. Isolation, structure, and partial synthesis of an active constituent of hashish. Journal of the American Chemical Society 1964; 86: 1646–1647.

[20] Lambert DM, Fowler CJ. The endocannabinoid system: Drug targets, lead compounds, and potential therapeutic applications. Journal of Medicinal Chemistry 2005; 48: 5059–5087.

[21] Pertwee RG. Cannabinoid receptors and pain. Progress in Neurobiology 2001; 63: 569–611.

[22] Pertwee RG, Ross R. Cannabinoid receptors and their ligands. Prostaglandins, Leukotrienes and Essential Fatty Acids (PLEFA) 2002; 66: 101–121.

[23] Adams R, Hunt M, Clark J. Structure of cannabidiol, a product isolated from the marihuana extract of Minnesota wild hemp. I. Journal of the American Chemical Society 1940; 62: 196–200.

[24] Mechoulam R, Shvo Y. Hashish – I: The structure of cannabidiol. Tetrahedron 1963; 19: 2073–2078.

[25] Mechoulam R, Gaoni Y. The absolute configuration of δ1-tetrahydrocannabinol, the major active constituent of hashish. Tetrahedron Letters 1967; 8: 1109–1111.

[26] Elliott J, DeJean D, Clifford T, Coyle D, Potter BK, Skidmore B, Alexander C, Repetski AE, Shukla V, McCoy B. Cannabis-based products for pediatric epilepsy: An updated systematic review. Seizure 2020; 75: 18–22.

[27] R de Mello Schier A, P de Oliveira Ribeiro N, S Coutinho D, Machado S, Arias-Carrión O, A Crippa J, W Zuardi A, E Nardi A, C Silva A. Antidepressant-like and anxiolytic-like effects of cannabidiol: A chemical compound of *Cannabis sativa*. CNS & Neurological Disorders-Drug Targets (Formerly Current Drug Targets-CNS & Neurological Disorders) 2014; 13: 953–960.

[28] US Food and Drug Administration (FDA). FDA approves first drug comprised of an active ingredient derived from marijuana to treat rare, severe forms of epilepsy. 2018.

[29] Moreno-Sanz G. Can you pass the acid test? critical review and novel therapeutic perspectives of Δ9-tetrahydrocannabinolic acid A. Cannabis and Cannabinoid Research 2016; 1: 124–130.

[30] Taschwer M, Schmid MG. Determination of the relative percentage distribution of THCA and Δ9-THC in herbal cannabis seized in Austria–Impact of different storage temperatures on stability. Forensic Science International 2015; 254: 167–171.

[31] Gaoni Y, Mechoulam R. Isolation and structure of. DELTA.+-tetrahydrocannabinol and other neutral cannabinoids from hashish. Journal of the American Chemical Society 1971; 93: 217–224.

[32] Mechoulam R, Braun P, Gaoni Y. Syntheses of. DELTA. 1-tetrahydrocannabinol and related cannabinoids. Journal of the American Chemical Society 1972; 94: 6159–6165.

[33] Adams R. Marihuana: Harvey lecture, February 19, 1942. Bulletin of the New York Academy of Medicine 1942; 18: 705.

[34] Adams R, Pease D, Cain C, Clark J. Structure of cannabidiol. VI. Isomerization of cannabidiol to tetrahydrocannabinol, a physiologically active product. Conversion of cannabidiol to cannabinol1. Journal of the American Chemical Society 1940; 62: 2402–2405.

[35] Adams R, Baker B. Structure of Cannabidiol. VII. A method of synthesis of a tetrahydrocannabinol which possesses Marihuana Activity1. Journal of the American Chemical Society 1940; 62: 2405–2408.

[36] Adams R, Loewe S, Pease D, Cain C, Wearn R, Baker R, Wolff H. Structure of cannabidiol. VIII. Position of the double bonds in cannabidiol. Marihuana activity of tetrahydrocannabinols. Journal of the American Chemical Society 1940; 62: 2566–2567.

[37] Gaoni Y, Mechoulam R. Hashish – VII: The isomerization of cannabidiol to tetrahydrocannabinols. Tetrahedron 1966; 22: 1481–1488.

[38] Petrzilka T, Sikemeier C. Über Inhaltsstoffe des Haschisch. 3., vorläufige Mitteilung. Umwandlung von (−)-Δ6, 1-3, 4-*trans*-Tetrahydrocannabinol in (−)-Δ1, 2-3, 4-*trans* Tetrahydrocannabinol. Helvetica Chimica Acta 1967; 50: 2111–2113.

[39] Petrzilka T, Haefliger W, Sikemeier C. Synthese von Haschisch-Inhaltsstoffen. 4. Mitteilung. Helvetica Chimica Acta 1969; 52: 1102–1134.

[40] Dethe DH, Erande RD, Mahapatra S, Das S, Kumar B. Protecting group free enantiospecific total syntheses of structurally diverse natural products of the tetrahydrocannabinoid family. Chemical Communications 2015; 51: 2871–2873.

[41] Stoss P, Merrath P. A useful approach towards Δ9-tetrahydrocannabinol. Synlett 1991; 1991: 553–554.

[42] Erman WF. Photochemical transformations of unsaturated bicyclic ketones. Verbenone and its photodynamic products of ultraviolet irradiation. Journal of the American Chemical Society 1967; 89: 3828–3841.

[43] Hurst J, Whitham G. 579. The photochemistry of verbenone. Journal of the Chemical Society (Resumed) 1960; 2864–2869.

[44] Razdan R, Handrick G, Dalzell H. A one-step synthesis of (-)-Δ1-Tetrahydrocannabinol from chrysanthenol. Experientia 1975; 31: 16–17.

[45] Evans DA, Shaughnessy EA, Barnes DM. Cationic bis (oxazoline) Cu (II) Lewis acid catalysts. Application to the asymmetric synthesis of ent-Δ1-tetrahydrocannabinol. Tetrahedron Letters 1997; 38: 3193–3194.

[46] Thomas A, Stevenson LA, Wease KN, Price MR, Baillie G, Ross RA, Pertwee RG. Evidence that the plant cannabinoid Δ9-tetrahydrocannabivarin is a cannabinoid CB1 and CB2 receptor antagonist. British Journal of Pharmacology 2005; 146: 917.

[47] Gill E, Paton WD, Pertwee RG. Preliminary experiments on the chemistry and pharmacology of cannabis. Nature 1970; 228: 134–136.

[48] Hill AJ, Williams CM, Whalley BJ, Stephens GJ. Phytocannabinoids as novel therapeutic agents in CNS disorders. Pharmacology and Therapeutics 2012; 133: 79–97.

[49] Hill A, Mercier M, Hill T, Glyn S, Jones N, Yamasaki Y, Futamura T, Duncan M, Stott C, Stephens G. Cannabidivarin is anticonvulsant in mouse and rat. British Journal of Pharmacology 2012; 167: 1629–1642.

[50] Hill T, Cascio MG, Romano B, Duncan M, Pertwee R, Williams C, Whalley B, Hill A. Cannabidivarin-rich cannabis extracts are anticonvulsant in mouse and rat via a CB1 receptor-independent mechanism. British Journal of Pharmacology 2013; 170: 679–692.

[51] Yamauchi T, Shoyama Y, Aramaki H, Azuma T, Nishioka I. Tetrahydrocannabinolic acid, a genuine substance of tetrahydrocannabinol. Chemical and Pharmaceutical Bulletin 1967; 15: 1075–1076.

[52] Kimura M, Okamoto K. Distribution of tetrahydrocannabinolic acid in fresh wild cannabis. Experientia 1970; 26: 819–820.

[53] Yoshinari S, Taro T, Kazuo K, Ayako T, Futoshi T, Shigeki A, Michael B, Yukihiro S, Satoshi M, Ryota K. Structure and function of Δ1-tetrahydrocannabinolic acid (THCA) synthase, the enzyme controlling the psychoactivity of *Cannabis sativa*. Journal of Molecular Biology 2012; 423: 96–105. DOI: https://doi.org/10.1016/j.jmb.2012.06.030.

[54] Carruthers W, Sutherland JK. Progress in organic chemistry. Springer, New York, 1973.

[55] Krejci Z, Šantavý F. Isolace dalš'ích látek z listí indického konopí *Cannabis sativa* L. Acta Univ Palacki Olomuc 1955; 6: 59–66.

[56] Petri G. *Cannabis sativa*: In vitro production of cannabinoids. In: Medicinal and aromatic plants I. Springer, Berlin, Heidelberg, 1988; 333–349.

[57] Garrett ER, Gouyette AJ, Roseboom H. Stability of tetrahydrocannabinols II. Journal of Pharmaceutical Sciences 1978; 67: 27–32. DOI: 10.1002/jps.2600670108.

[58] Adams R, Baker BR, Wearn RB. Structure of cannabinol. III. Synthesis of cannabinol, 1-hydroxy-3-n-amyl-6,6,9-trimethyl-6-dibenzopyran1. Journal of the American Chemical Society 1940; 62: 2204–2207. DOI: 10.1021/ja01865a083.

[59] Karniol IG, Shirakawa I, Takahashi RN, Knobel E, Musty RE. Effects of Δ9-tetrahydrocannabinol and cannabinol in man. Pharmacology 1975; 13: 502–512.

[60] Farrimond JA, Whalley BJ, Williams CM. Cannabinol and cannabidiol exert opposing effects on rat feeding patterns. Psychopharmacology 2012; 223: 117–129.

[61] Caprioglio D, Mattoteia D, Minassi A, Pollastro F, Lopatriello A, Muñoz E, Taglialatela-Scafati O, Appendino G. One-pot total synthesis of cannabinol via iodine-mediated deconstructive annulation. Organic Letters 2019; 21: 6122–6125.

[62] Turner CE, Elsohly MA. Constituents of *Cannabis sativa* L. XVI. A possible decomposition pathway of Δ9-tetrahydrocannabinol to cannabinol. Journal of Heterocyclic Chemistry 1979; 16: 1667–1668.

[63] Bercht C, Lousberg R, Küppers F, Salemink C, Vree T, Van Rossum J. Cannabis: VII. Identification of cannabinol methyl ether from hashish. Journal of Chromatography A 1973; 81: 163–166.

[64] Turner CE, Elsohly MA, Boeren EG. Constituents of *Cannabis sativa* L. XVII. A review of the natural constituents. Journal of Natural Products 1980; 43: 169–234.

[65] Uliss DB, Razdan RK, Dalzell HC. Stereospecific intramolecular epoxide cleavage by phenolate anion. Synthesis of novel and biologically active cannabinoids. Journal of the American Chemical Society 1974; 96: 7372–7374.

[66] Gaoni Y, Mechoulam R. Cannabichromene, a new active principle in hashish. Chemical Communications (London), 1966; 1:20–21.

[67] Claussen U, Von Spulak F, Korte F. Chemical classification of plants. XXXI. Hashish. 10. Cannabichromene, a new hashish component. Tetrahedron 1966; 22: 1477–1479.

[68] Anderson LL, Ametovski A, Lin Luo J, Everett-Morgan D, McGregor IS, Banister SD, Arnold JC. Cannabichromene, related phytocannabinoids, and 5-fluoro-cannabichromene have anticonvulsant properties in a mouse model of Dravet Syndrome. ACS Chemical Neuroscience 2021; 12: 330–339.

[69] Wirth PW, Watson ES, ElSohly M, Turner CE, Murphy JC. Anti-inflammatory properties of cannabichromene. Life Sciences 1980; 26: 1991–1995.

[70] Claussen U, Spulak F, Korte F. Zur chemischen klassifizierung von pflanzen – XXXI, haschisch – X: Cannabichromen, ein neuer haschisch-inhalts-stoff. Tetrahedron 1966; 22: 1477–1479.

[71] Brown N, Harvey D. In vitro metabolism of cannabichromene in seven common laboratory animals. Drug Metabolism and Disposition 1990; 18: 1065–1070.

[72] Davis WM, Hatoum NS. Neurobehavioral actions of cannabichromene and interactions with delta 9-tetrahydrocannabinol. General Pharmacology 1983; 14: 247–252.

[73] Turner CE, Elsohly MA. Biological activity of cannabichromene, its homologs and isomers. The Journal of Clinical Pharmacology 1981; 21: 283S–291S.

[74] Wirth PW, ES W, MA E, JC M. Anti-inflammatory activity of cannabichromene homologs. 1980.

[75] Mechoulam R, Yagnitinsky B, Gaoni Y. Hashish. XII. Stereoelectronic factor in the chloranil dehydrogenation of cannabinoids. Total synthesis of dl-cannabichromene. Journal of the American Chemical Society 1968; 90: 2418–2420.

[76] ElSohly MA, Boeren EG, Turner CE. Constituents of *Cannabis sativa* L. An improved method for the synthesis of dl-cannabichromene. Journal of Heterocyclic Chemistry 1978; 15: 699–700.

[77] Cardillo G, Cricchio R, Merlini L. Synthesis of d, l-cannabichromene, franklinone and other natural chromenes. Tetrahedron 1968; 24: 4825–4831.

[78] Kirby K. Recent developments in the chemistry of natural phenolic compounds: Edited WD Ollis, Pergamon Press, 1961, 70s. Elsevier, New York, 1962.

[79] Campbell I, Calzadilla C, McCorkindale N. Some new metabolites related to mycophenolic acid. Tetrahedron Letters 1966; 7: 5107–5111.

[80] Korte F, Sieper H. Zur chemischen klassifizierung von pflanzen: XXIV. Untersuchung von Haschisch-Inhaltsstoffen durch Dünnschichtchromatographie. Journal of Chromatography A 1964; 13: 90–98.

[81] Mechoulam R, Gaoni Y. Recent advances in the chemistry of hashish. Fortschritte der Chemie Organischer Naturstoffe/Progress in the Chemistry of Organic Natural Products/Progrès Dans la Chimie Des Substances Organiques Naturelles 1967; 25: 175–213.

[82] Bercht CL, Lousberg RJ, Küppers FJ, Salemink CA. Cannabicitran: A new naturally occurring tetracyclic diether from lebanese *Cannabis sativa*. Phytochemistry 1974;13: 619–21.

[83] Elsohly MA, Harland EC, Benigni DA, Waller CW. Cannabinoids in glaucoma II: The effect of different cannabinoids on intraocular pressure of the rabbit. Current Eye Research 1984; 3: 841–850.

Mohamed M. Radwan, Amira S. Wanas, Suman Chandra
and Mahmoud A. ElSohly*

6 Biosynthesis of cannabis constituents

Abstract: Cannabinoids, non-cannabinoid phenols, flavonoids, and terpenes are the major secondary metabolites of *Cannabis sativa*. Many enzymes and intermediates are involved in the process of biosynthesis of these metabolites. Oxidases, transferases, reductases, and isomerases are examples of these enzymes which are responsible for various chemical conversions of intermediates inside the plant's tissues. In this chapter, the biosynthesis of cannabinoids, non-cannabinoid phenols, flavonoids, and terpenes is discussed.

Keywords: biosynthesis, enzymes, cannabinoids, non-cannabinoid phenols, flavonoids, terpenes

6.1 Introduction

More than 550 metabolites have been identified in cannabis [1], which can be categorized into primary and secondary metabolites. Primary metabolites are essential for plants and all organisms' life to produce energy. They provide the key intermediates for the biosynthesis of secondary metabolites. Proteins, fats, carbohydrates, nucleic acids, and steroids are examples of primary metabolites, which are produced under all conditions by primary metabolic pathways such as Krebs cycle (the tricarboxylic acid cycle), fatty acid β-oxidation, photosynthesis, and glycolysis [2]. Secondary metabolites are produced under certain conditions to act as plant defense against insects and diseases or to provide aroma to attract insects for pollination or as coloring agents. Most of the pharmacological activities are due to secondary metabolites. Cannabinoids, non-cannabinoid phenols, flavonoids, and terpenes are the major secondary metabolites in cannabis [1], which are biosynthesized by secondary metabolic pathways such as acetate, mevalonate phosphate (MVP), methylerythritol phosphate (MEP), and shikimate pathways. Biosynthesis is an enzyme-catalyzed process that occurs via primary and secondary metabolic pathways, where substrates are converted into more complex products in living organisms.

---

*Corresponding author: Mahmoud A. ElSohly**, National Center for Natural Products Research and Department of Pharmaceutics and Drug Delivery School of Pharmacy, University of Mississippi, University, MS 38677, USA, e-mail: melsohly@olemiss.edu
Mohamed M. Radwan, Amira S. Wanas, Suman Chandra, National Center for Natural Products Research, School of Pharmacy, University of Mississippi, University, MS 38677, USA

https://doi.org/10.1515/9783110718362-006

6.2 Biosynthesis of cannabinoids

Cannabinoids are biosynthesized mainly in the glandular trichome of *Cannabis sativa* but they are also detected in the stems and leaves [3, 4]. All cannabinoid structures have monoterpene residues (C_{10}) connected to a phenolic ring that carries an alkyl residue (mainly C_5 and C_3, and to a lesser extent C_4, C_1, and C_7) as shown in Figure 6.1.

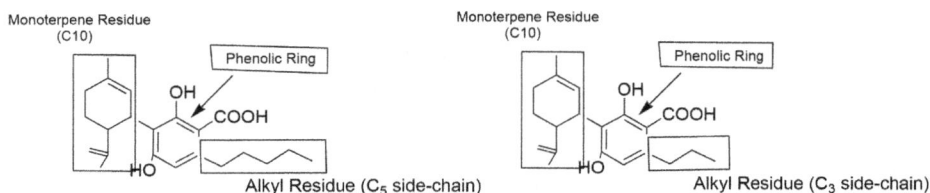

Figure 6.1: Chemical structures of the components of cannabinoid acids.

Steps for the biosynthesis of cannabinoids with C_5 side chain are:
1. Biosynthesis of olivetolic acid (**OA**)
2. Biosynthesis of geranyl pyrophosphate (**GPP**) or neryl pyrophosphate (**NPP**)
3. Biosynthesis of cannabigerolic acid (**CBGA**)
4. Biosynthesis of tetrahydrocannabidiolic acid (**THCA**), cannabidiolic acid (**CBDA**), and cannabichrominic acid (**CBCA**)
5. Formation of neutral cannabinoids

6.2.1 Biosynthesis of olivetolic acid

OA is a dihydroxy phenolic acid which constitutes the phenolic ring/C_5 side-chain part of the cannabinoid' structures. It is formed by the condensation of one molecule of **hexanoyl-CoA** and three molecules of **malonyl-CoA** by *polyketide synthase* (PKS) enzyme (Figure 6.2).

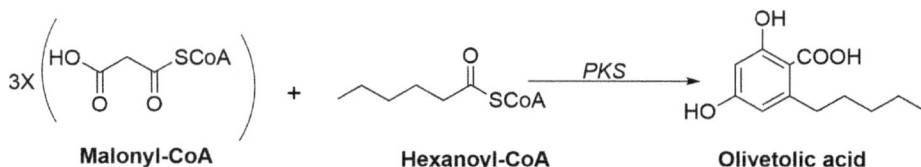

Figure 6.2: Formation of olivetolic acid.

Both **hexanoyl-CoA** and **malonyl-CoA** are derived from the acetate metabolic pathway [2].

6.2.1.1 Formation of malonyl-CoA and hexanoyl-CoA

Malonyl-CoA originates from acetyl-CoA. Acetyl-CoA is the key intermediate in many pathways. It is produced from the glycolysis pathway. *Acetyl-CoA carboxylase* enzyme catalyzes the conversion of acetyl-CoA to **malonyl-CoA** (Figure 6.3).

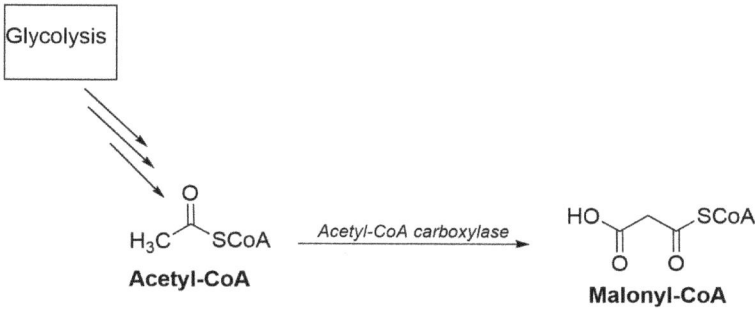

Figure 6.3: Formation of malonyl-CoA.

Hexanoyl-CoA is formed in cannabis trichomes from hexanoic acid by the *hexanoyl-CoA synthetase* enzyme (Figure 6.4), and hexanoic acid is biosynthesized from the fatty acid biosynthesis pathway [5].

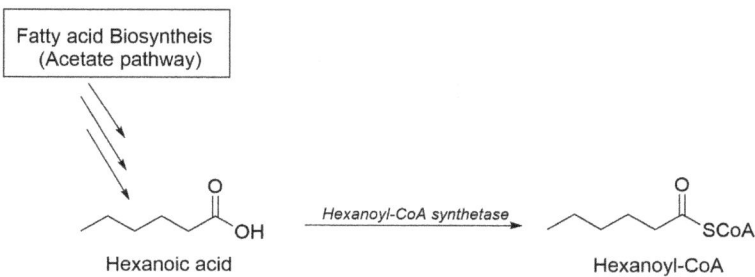

Figure 6.4: Formation of hexanoyl-CoA.

6.2.2 Biosynthesis of GPP

GPP or geranyl diphosphate is derived from **MVP** and **MEP** pathways. Both pathways form two reactive hemiterpene (C_5) intermediates named isopentenyl pyrophosphate or isopentenyl diphosphate (**IPP**) and dimethylallyl pyrophosphate or dimethylallyl diphosphate (**DMAPP**) (Figure 6.5). These two intermediates are the precursors of all terpenes (monoterpenes (C_{10}), sesquiterpenes (C_{15}), diterpenes (C_{20}), sesterterpenes (C_{25}), and triterpenes (C_{30}).

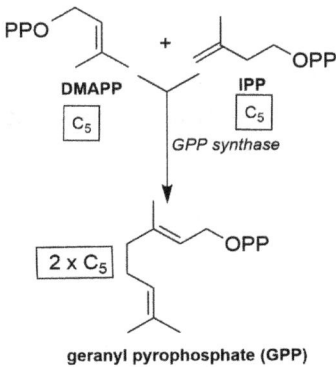

geranyl pyrophosphate (GPP) **Figure 6.5:** Formation of geranyl pyrophosphate (**GPP**).

6.2.3 Biosynthesis of CBGA

CBGA is the precursor of *n*-pentyl cannabinoid acids or cannabinoid acids with C_5 side chain (**THCA, CBDA**, and **CBCA**). These three acids are the parent cannabinoids for all C_5 acid cannabinoids and C_5 neutral cannabinoids [6]. Neutral cannabinoids are derived from nonenzymatic decarboxylation of their corresponding acid cannabinoids. **CBGA** is formed by the condensation of **OA** and **GPP** by *geranylpyrophosphate: olivetolate geranyltransferase* (GOT) enzyme or CBGA synthase enzyme (Figure 6.6).

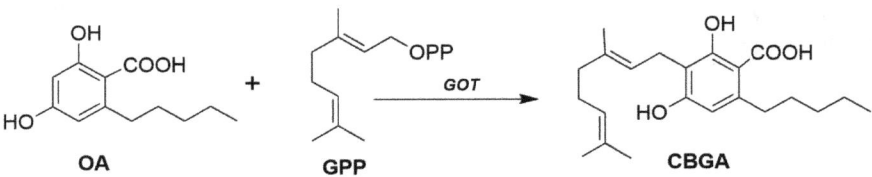

Figure 6.6: Biosynthesis of cannabigerolic acid (**CBGA**).

6.2.4 Biosynthesis of THCA, CBDA, and CBCA

Three enzymes are responsible for the conversion of **CBGA** to **THCA, CBDA**, and **CBCA**. Oxidative cyclization of **CBGA** by *THCA synthase* (THCAS) enzyme resulted in the formation of **THCA**, while *CBDA synthase* (CBDAS) catalyzes the oxidocyclization of **CBGA** to **CBDA**. Both THCAS and CBDAS are expressed and stored in the glandular trichomes of cannabis [7, 8]. It is also proven that **CBGA** is cyclized to **CBCA** by *CBCA synthase* [9] as demonstrated in Figure 6.7.

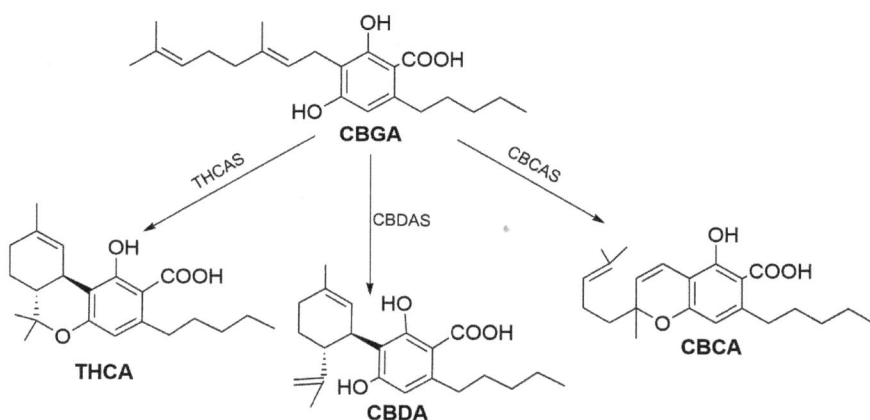

Figure 6.7: Biosynthesis of THCA, CBDA, and CBCA from CBGA.

6.2.5 Formation of neutral cannabinoids

Most cannabinoids are formed mainly due to nonenzymatic transformation or degradation of both acidic and neutral cannabinoids as a result of the effects of light (UV irradiation), heat, and oxidation. Therefore, it is still doubtful if all these forms are present as natural constituents or just artifacts originating during storage and/or sample preparation [10].

Neutral cannabinoids are derived from their corresponding acid cannabinoids by nonenzymatic decarboxylation during the drying of the cannabis plants or during extraction. Decarboxylation certainly occurs during smoking. **CBGA**, Δ^9**-THCA**, **CBDA**, and **CBCA** are decarboxylated to their corresponding neutral cannabinoids, **CBG**, Δ^9**-THC**, **CBD**, and **CBC** (Figure 6.8).

Cannabinol (**CBN**) is an oxidized product resulting from the oxidation (aromatization) of Δ^9**-THC** upon long storage and exposure to light [11]. Δ^9**-THCA** and **CBDA** are also oxidized to cannabinolic acid (**CBNA**) and cannabinodiol (**CBND**), respectively (Figure 6.8). Since **CBD** is a stable cannabinoid, **CBND** may be formed as a result of enzymatic activity.

Δ^8**-THC** resulted from the nonenzymatic isomerization of Δ^9**-THC** (Figure 6.9). It is thermodynamically more stable than Δ^9**-THC** and does not oxidize nearly as easy as Δ^9**-THC** to **CBN** upon storage. Δ^8**-THC** might be an artifact of extraction and/or analysis process.

Cannabicyclolic acid (**CBLA**) is an artifact originated from natural irradiation (UV light) of **CBCA**, and it is decarboxylated to cannabicyclol (**CBL**) by heat. **CBL** was also originated from **CBC** after UV irradiation [12] (Figure 6.10).

The formation of cannabielsoic acid (**CBEA**) results from the photooxidative cyclization of **CBDA** [13]. Its neutral form, cannabielsoin (**CBE**), was found to be the main product of the pyrolysis of **CBD** in air [14]. **CBE** was also identified as the main **CBD** metabolite in guinea pigs, which is possibly formed through epoxy CBD intermediate by *cytochrome P-450* [15, 16]. Formation of **CBEA** and **CBE** is shown in Figure 6.11.

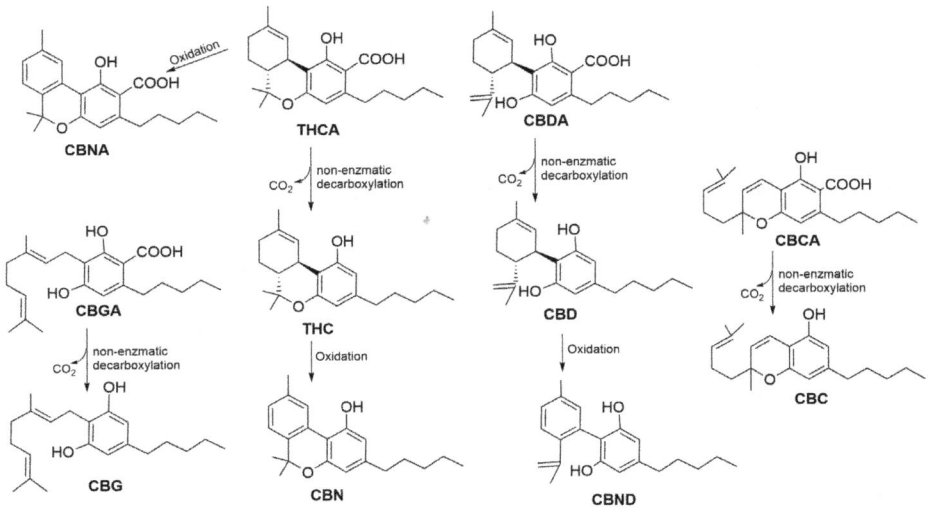

Figure 6.8: Decarboxylation and oxidation of cannabinoids.

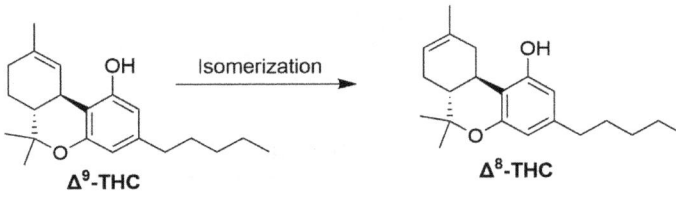

Figure 6.9: Isomerization of Δ^9-THC to Δ^8-THC.

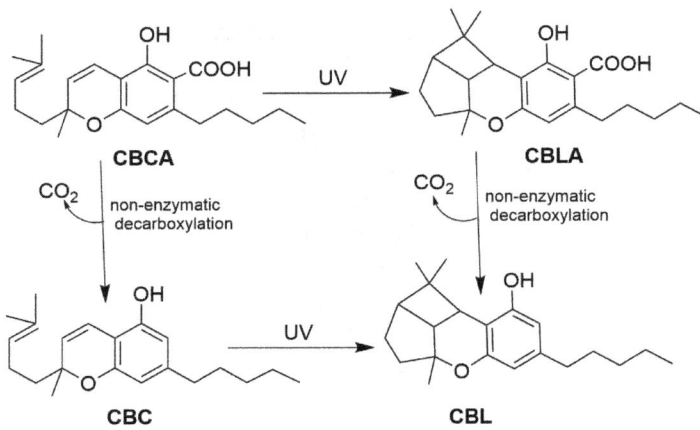

Figure 6.10: Formation of CBLA and CBL.

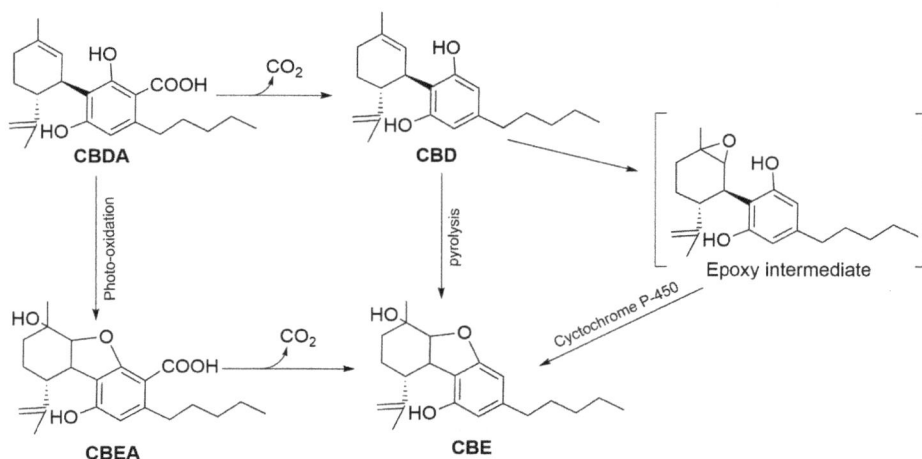

Figure 6.11: Formation of CBEA and CBE.

6.2.6 Biosynthesis of propyl cannabinoids

Cannabigerovarinic acid (**CBGVA**) is the precursor for the biosynthesis of propyl cannabinoids or cannabinoids with C_3 side chain. Formation of propyl cannabinoids is not the result of shortening of the C_5 side chain of the n-pentyl cannabinoids [17, 18]. **Divarinolic acid** is the parent compound in the biosynthesis of **CBGVA**, which is formed by condensation of one molecule of n-**butyl-CoA** and three molecules of **malonyl-CoA** by PKS. Alkylation of **divarinolic acid** by **GPP** in the presence of *GOT* enzyme resulted in the formation of **CBGVA** (Figure 6.12).

The enzymes that are responsible for the biosynthesis of the acid cannabinoids with C_5 side chain are also responsible for the formation of the cannabinoid acids with C_3 side chain. These enzymes are not selective for the length of the side chain. The biosynthesized cannabinoid acids are called tetrahydrocannabivarinic acid (**THCVA**), cannabidivarinic acid (**CBDVA**), and cannabichromevarinic acid (**CBCVA**) as shown in Figure 6.12.

Like n-pentyl cannabinoids, most of n-propyl neutral cannabinoids are formed as a result of nonenzymatic decarboxylation, oxidation, and cyclization of propyl cannabinoid acids (Figure 6.13). Cannabigerovarin (**CBGV**), tetrahydrocannabivarin (**THCV**), cannabidivarin (**CBDV**), cannabichromevarin (**CBCV**), cannabivarin (**CBV**), and cannabicyclovarin (**CBLV**) are the major neutral cannabinoids formed after decarboxylation of their corresponding cannabinoid acids (Figure 6.13). Cannabivarinic acid (**CBVA**), **CBV**, cannabinodivarin (**CBVD**) artifacts formed by oxidation of **THCVA**, **THCV**, and **CBDV**, respectively, upon long storage and heat. Cannabicyclovarinic acid (**CBLVA**) is produced from **CBCA** by natural irradiation (UV light), and it is decarboxylated to cannabicyclovarin (**CBLV**) by heat. **CBLV** was also formed as a result of UV irradiation of **CBCV**.

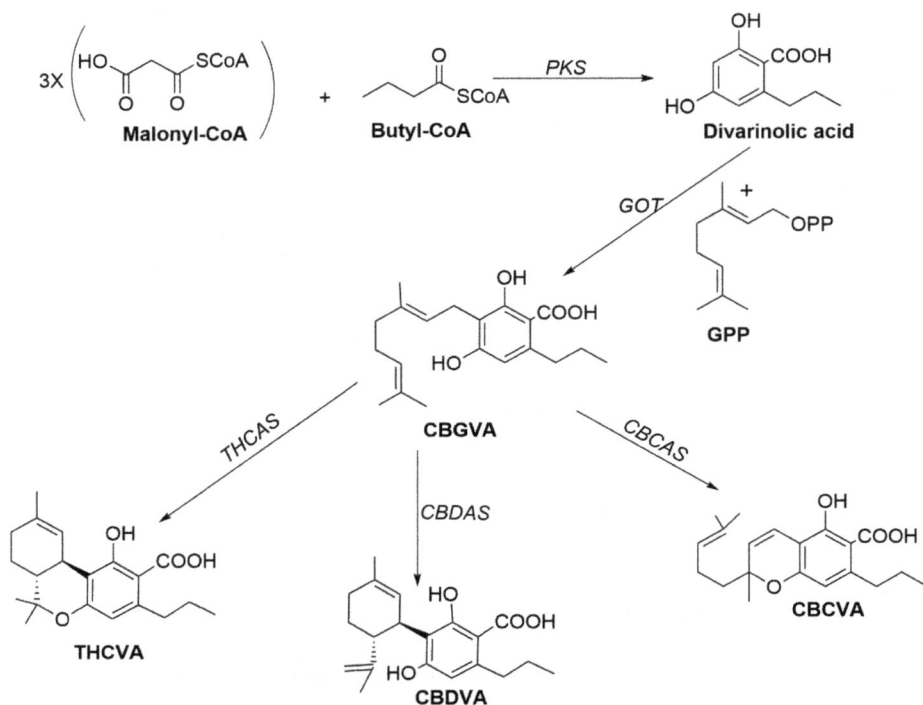

Figure 6.12: Biosynthesis of propyl cannabinoid acids.

6.3 Biosynthesis of non-cannabinoid constituents

The major non-cannabinoid constituents of cannabis include non-cannabinoid phenols, flavonoids, and terpenes. Forty-two non-cannabinoid phenols have been identified from cannabis of different chemical classes, including dihydrostilbenes, cannabispirans and dihydrophenanthrenes [1].

6.3.1 Biosynthesis of non-cannabinoid phenols

6.3.1.1 Biosynthesis of dihydrostilbenes

Phenyl alanine and **malonyl-CoA** are the building blocks in the biosynthesis of dihydrostilbenes. **Phenyl alanine** is derived from the shikimate pathway, while **malonyl-CoA** is biosynthesized from the acetate pathway [19]. **Phenyl alanine** is converted to dihydro-*p*-coumaroyl-CoA that condensed with three molecules of malonyl-CoA by *bibenzyl synthase* (BBS) enzyme to produce **dihydroresveratrol** (Figure 6.14).

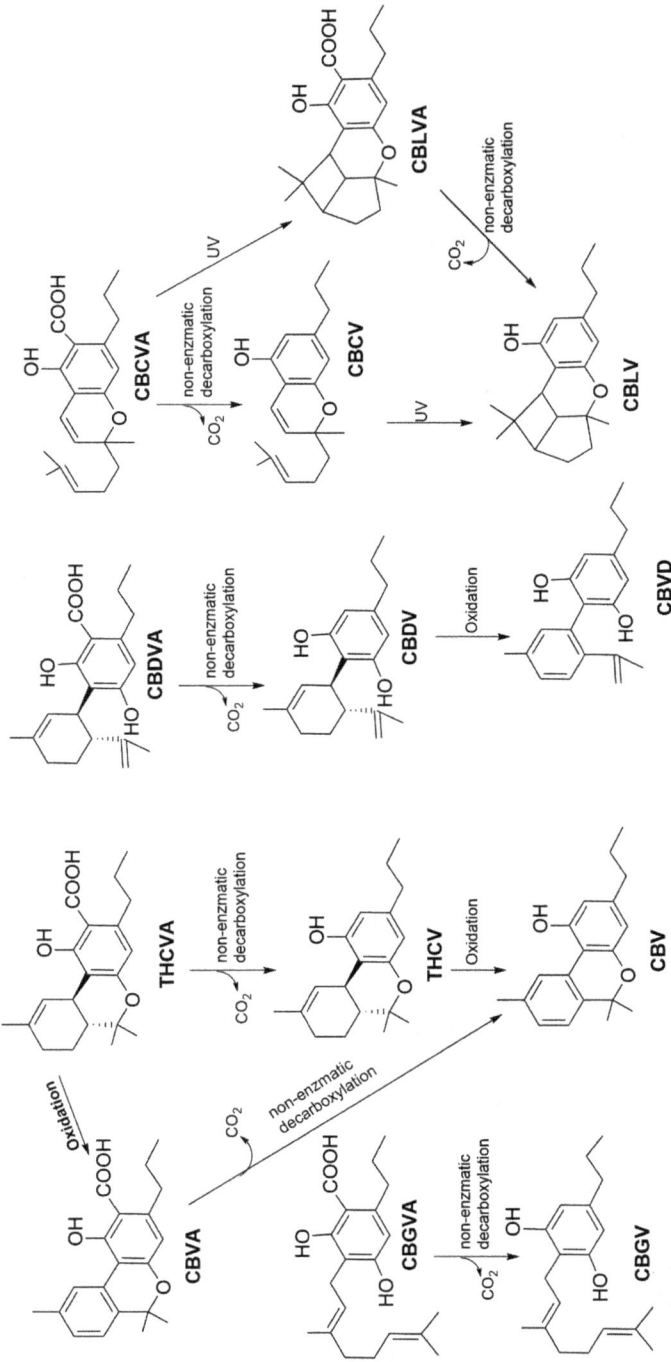

Figure 6.13: Formation of neutral propyl cannabinoids.

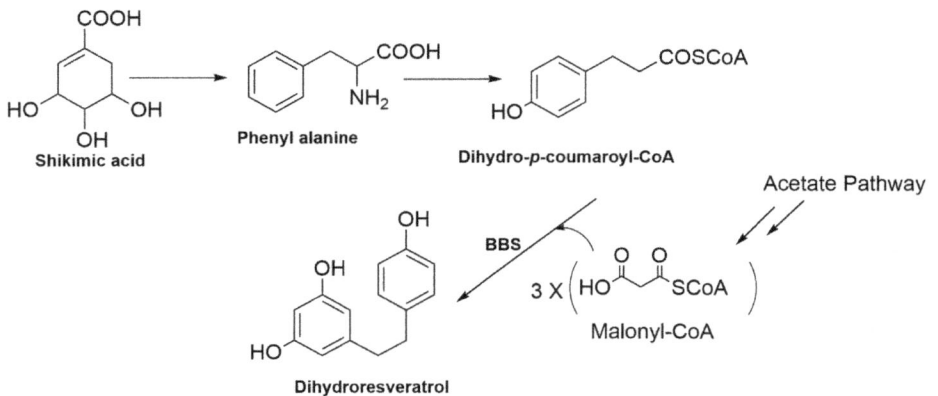

Figure 6.14: Biosynthesis of dihydroresveratrol.

Modification reactions like methylation and prenylation of dihydroresveratrol form most of dihydrostilbenes (Figure 6.15). Methylation of **dihydroresveratrol** by *O-methyl transferase* (OMT) forms 3,4′-dihydroxy-5-methoxybibenzyl, which is subjected to prenylation to form 3,4′-dihydroxy-5-methoxy-3′-prenylbibenzyl. *Prenyltransferase* is the enzyme responsible for prenylation. Some prenylated and methoxylated dihydrostilbene derivatives are also shown in Figure 6.15.

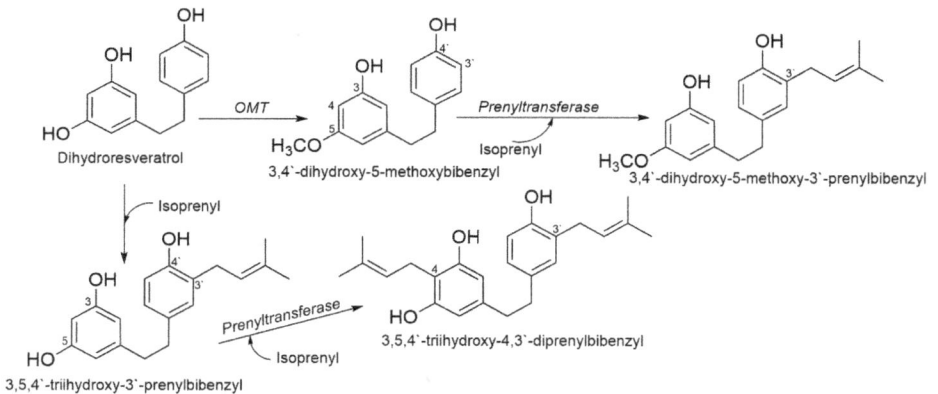

Figure 6.15: Biosynthesis of prenylated and methoxylated dihydroresveratrol.

6.3.1.1.1 Biosynthesis of canniprene

Canniprene is a prenylated dimethoxy dihydrostilbene that was isolated from the leaves of high THC cannabis biomass of Thailand origin [20]. Similar to dihydroresveratrol, **canniprene** is biosynthesized from **phenyl alanine** as shown in Figure 6.16. **Dihydrocaffeoyl-CoA** is condensed with three molecules of **malonyl-CoA** by BBS enzyme to form 3,5,3′,4′-tetrahydroxybibenzyl [21], which is methylated by OMT enzyme to 3,3′-dihydroxy-5,4′-dimethoxy bibenzyl [21] and then prenylated by *prenyltransferase* to form **canniprene** (Figure 6.16).

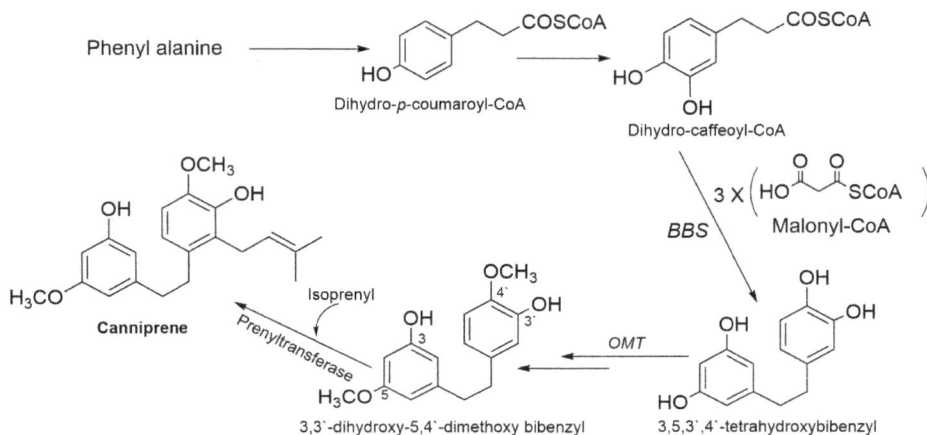

Figure 6.16: Biosynthesis of canniprene.

6.3.1.2 Biosynthesis of cannabispirans

Cannabispirans are one of the important chemical classes in cannabis, which are derived from dihydrostilbenes and are considered an intermediate step between the biosynthesis of dihydrostilbenes and dihydrophenanthrenes. The biosynthesis of cannabispirans starts with one electron oxidation of dihydrostilbenes followed by *p–o* or *p–p* coupling [22]. **Cannabispirone** is an example of cannabispirans which was isolated from the cultivated hemp of South African origin [23]. It is biosynthesized from 3,4′-dihydroxy-5-methoxybibenzyl after oxidation to give diradical intermediate that couples in *p–o* pattern to form **cannabispiradienone**. Successive reduction of **cannabispiradienone** results in the formation of **cannabispirenone A**, **cannabispirone**, and **cannabispiranol**. 3,4′-dihydroxy-5-methoxybibenzyl is also the precursor of **cannabispirenone B** after *p–p* coupling of its oxidizing diradical intermediate followed by reduction (Figure 6.17). **Cannabispiradienone** could be prenylated by the *prenyltransferase* enzyme to produce **prenylspirodinone** [24]. The biogenesis of some examples of **cannabispirans** is shown in Figure 6.17.

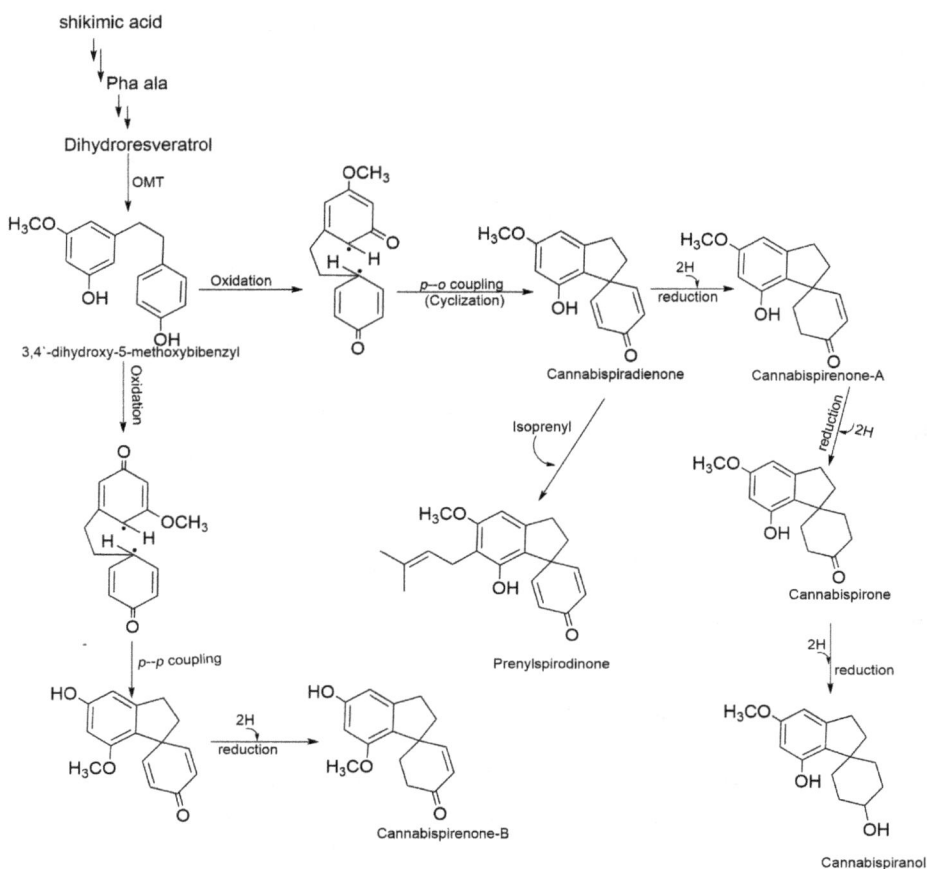

Figure 6.17: Biosynthesis of cannabispirans.

6.3.1.3 Biosynthesis of dihydrophenanthrenes

Dihydrophenanthrenes are biosynthesized from **bibenzyls** via cyclization. Like dihydrostilbenes, their biosynthesis is derived from **phenylalanine**. Methylation of bibenzyls by *OMT* enzyme is essential before cyclization to dihydrophenanthrenes [25]. **Cannithrenes 1** and **2** are 9,10-dihydrophenanthrenes isolated from Thailand's high THC *C. sativa* [22, 26]. **9,10-Dihydrophenanthrene** could also originate from spirans via dienone–phenol rearrangement on heating or under acidic conditions [22]. The biosynthesis of **cannithrene 1** and **cannithrene 2** is illustrated in Figure 6.18.

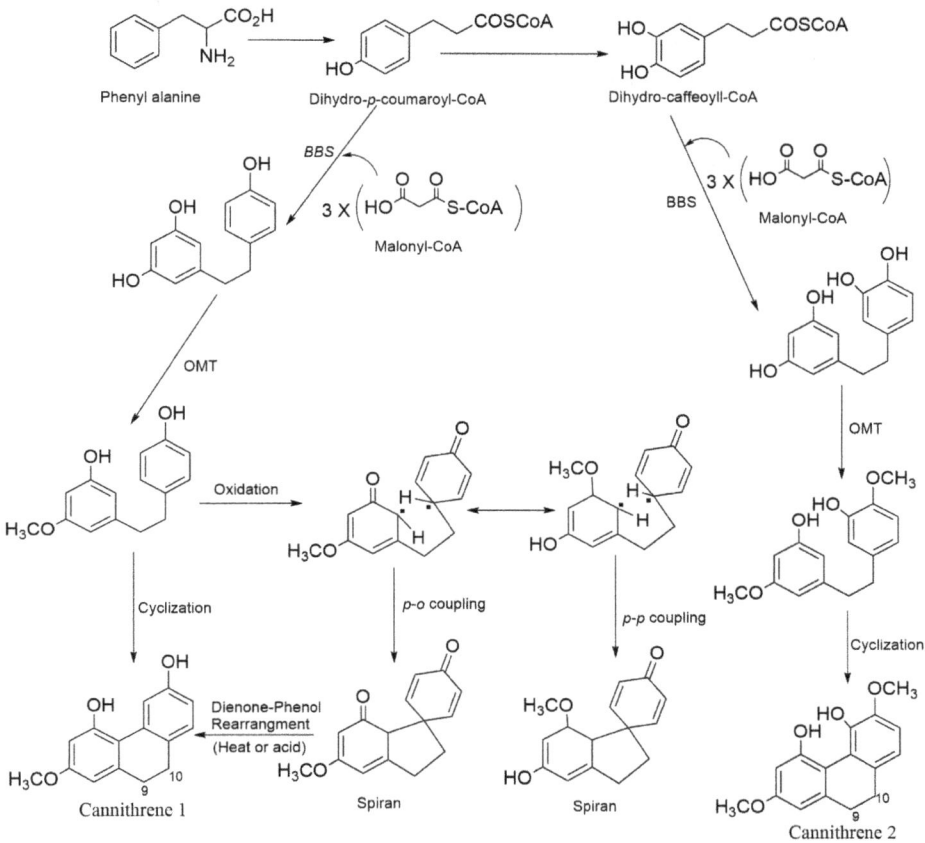

Figure 6.18: Biosynthesis of cannithrenes 1 and 2.

6.3.2 Biosynthesis of flavonoids

More than 30 flavonoids have been identified from cannabis belonging to 7 chemical subclasses that can be hydroxylated, methylated, glycosylated, or prenylated. The seven chemical structures of flavonoids are orientin, vitexin, isovitexin, apigenin, luteolin, kaempferol, and quercetin [1]. **Naringenin** aglycone is the precursor for the biosynthesis of all cannabis flavonoids.

6.3.2.1 Biosynthesis of naringenin

The first step in the biosynthesis of **naringenin** is the formation of *p*-cinnamic acid from **phenyl alanine** (originating from the shikimate pathway) by *phenylalanine ammonia* *lyase* enzyme (PAL), then the hydroxylation of *p*-cinnamic acid to *p*-coumaric acid by *cinnamate-4-hydroxylase* (C4H), which is activated to *p*-coumaroyl-CoA by *4-coumarate:CoA ligase* (4CL). **p-Coumaroyl-CoA** is condensed with three molecules of **malonyl-CoA** (originating from the carboxylation of acetyl-CoA) by *chalcone synthase* (CHS) enzyme to yield naringenin chalcone, which isomerizes to **naringenin** by *chalcone isomerase* (CHI) enzyme (Figure 6.19).

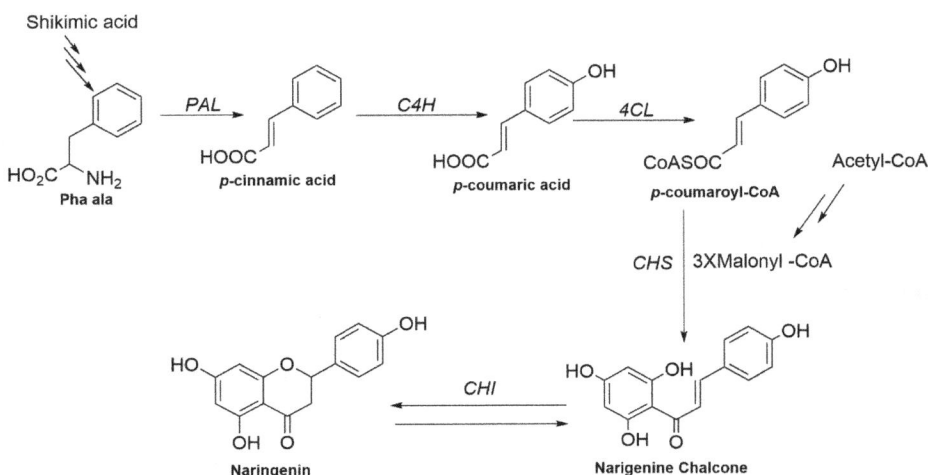

Figure 6.19: Biosynthesis of naringenin.

6.3.2.2 Biosynthesis of different cannabis flavonoids from naringenin

Naringenin is a flavanone aglycone consisting of three rings A, B, and C. Double bond is formed in ring C between C_2 and C_3 by *flavone synthase* (FNS) to form **apigenin**. Glycosylation of apigenin ring A at C_6 and C_8 by *UDP-glycosyltransferase* (UGT) gives **isovitexin** and **vitexin**, respectively. **Luteolin** is formed by the hydroxylation of ring B of **apigenin** at position 3′ by *flavonoid-3′-hydroxylase* (F3′H). Modification of **luteolin** by UGT enzyme yields the flavonoid glycoside **orientin**. Hydroxylation of **naringenin** at C_3 of ring C by *flavonoid-3-hydroxylase* (F3H) yields dihydrokaempferol which is converted to **kaempferol** by *flavonol synthase* (FLS). **Quercetin** is formed by the hydroxylation of **kaempferol** at $C_{3′}$ by F3′H. The biosynthetic pathways of flavonoids are shown in Figure 6.20.

6.3.2.3 Biosynthesis of cannflavins A, B, and C

Cannflavins A, B, and **C** are prenylated flavonoids isolated from the ethanolic extract of *C. sativa* [1, 27, 28]. They are biosynthesized from **luteolin**, which is at first methylated by

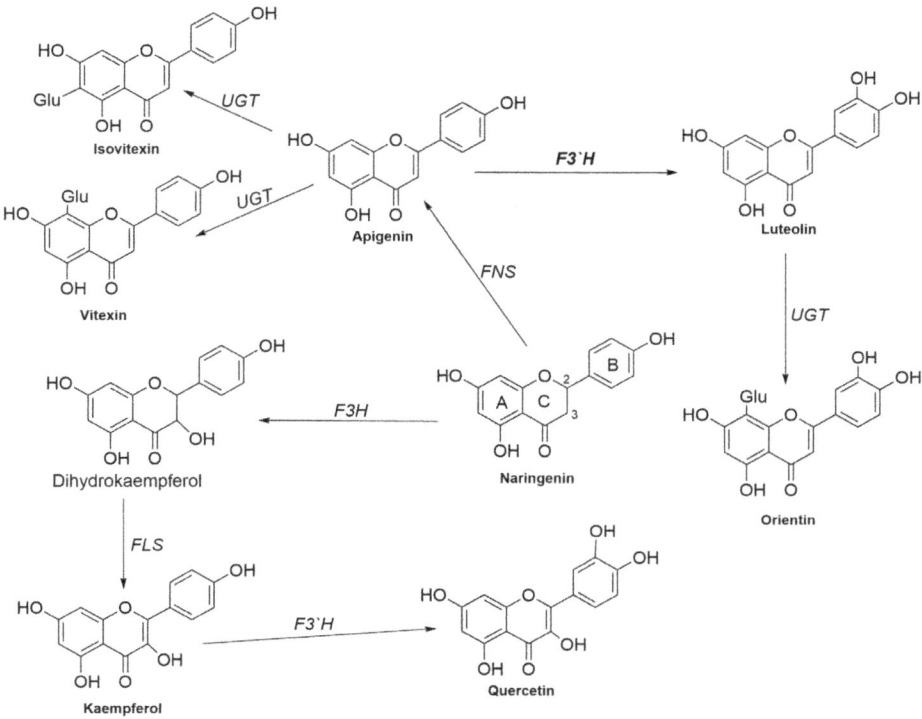

Figure 6.20: Biosynthesis of different cannabis flavonoids.

OMT to **chrysoeriol** and then prenylation of ring A at C_6 or C_8 to form cannflavins. Prenylation occurs by *prenyltransferase* enzyme which is able to introduce isoprenyl at C_6 to form **cannflavin B** or geranyl at C_6 to yield **cannflavin A** or geranyl at C_8 to form **cannflavin C** (Figure 6.21). The biosynthesis of **cannflavins A** and **B** was proven using a combination of phylogenomic and biochemical approaches [29].

6.3.3 Biosynthesis of terpenes

Terpenes are the second largest group of metabolites in cannabis after cannabinoids with more than 100 reported compounds [1]. These compounds are responsible for the characteristic aroma and the flavor of the plant. Terpenes are biosynthesized from both MVP and MEP pathways [2]. Both pathways form two reactive **hemiterpene (C_5)** intermediates named **isopentenyl pyrophosphate (IPP)** and **dimethylallyl pyrophosphate (DMAPP)** (Figure 6.22). These two intermediates are condensed by *geranyl pyrophosphate synthase* (GPP synthase) to produce **GPP**. **NPP** is the isomer of

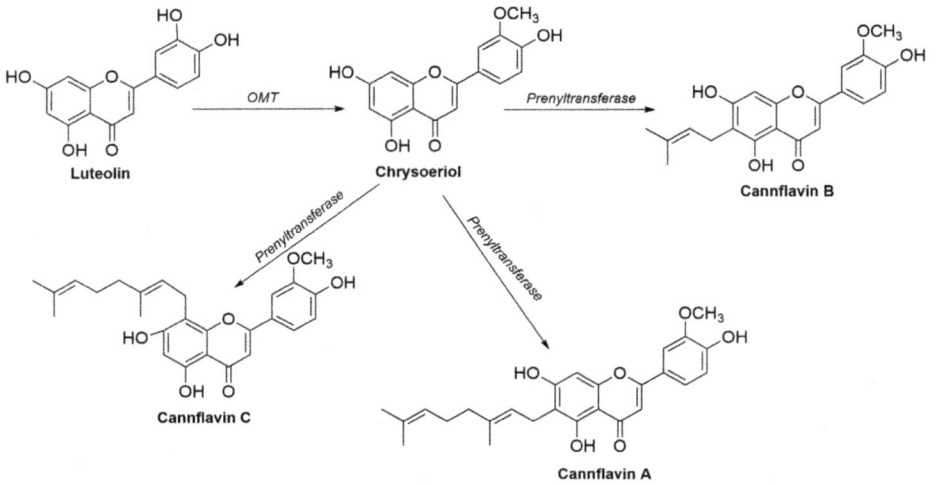

Figure 6.21: Biosynthesis of cannflavins A, B, and C.

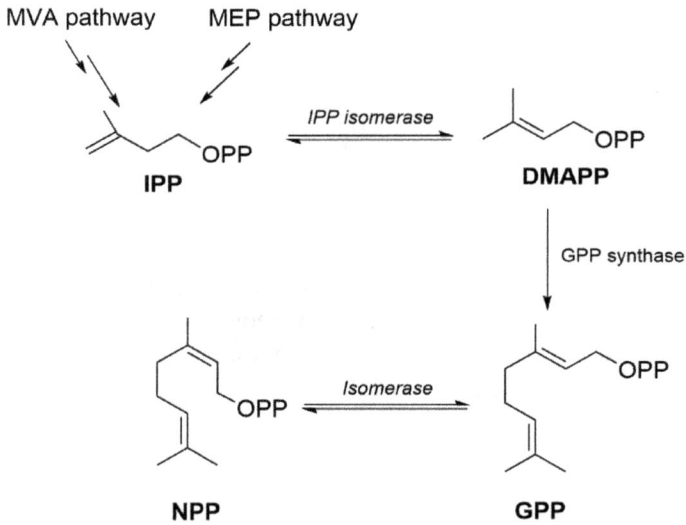

Figure 6.22: Biosynthesis of GPP and NPP.

GPP. The intermediates **GPP** and **NPP** are the precursors of all the immediate precursors for all cannabis monoterpenes (Figure 6.22).

6.3.3.1 Biosynthesis of monoterpenes

More than 60 monoterpenes are detected in cannabis [1], which can be classified into acyclic (linear), monocyclic, and bicyclic monoterpenes. **Geraniol, citronellol, linalool**, and **β-myrcene** are examples of the major acyclic cannabis monoterpenes which originate from **GPP** and **NPP**. *Geraniol synthase* enzyme converts **GPP** into the acyclic monoterpene **geraniol** which undergoes reduction of the allylic alcohol double bond to produce **citronellol** (Figure 6.23).

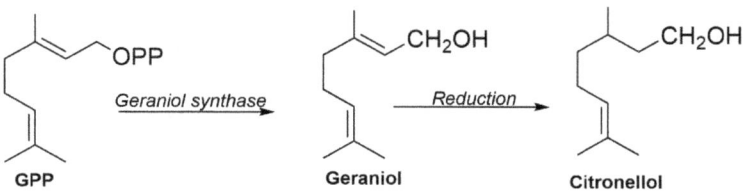

Figure 6.23: Biosynthesis of geraniol and citronellol.

Linalool synthase acts on **neryl cation intermediate** that originates from **NPP** to form **linalool**. *β*-**Myrcene** is one of the major linear unsaturated monoterpenes biosynthesized by the *myrcene synthase* enzyme after losing one hydrogen from the **neryl cation intermediate** to form an extra double bond (Figure 6.24).

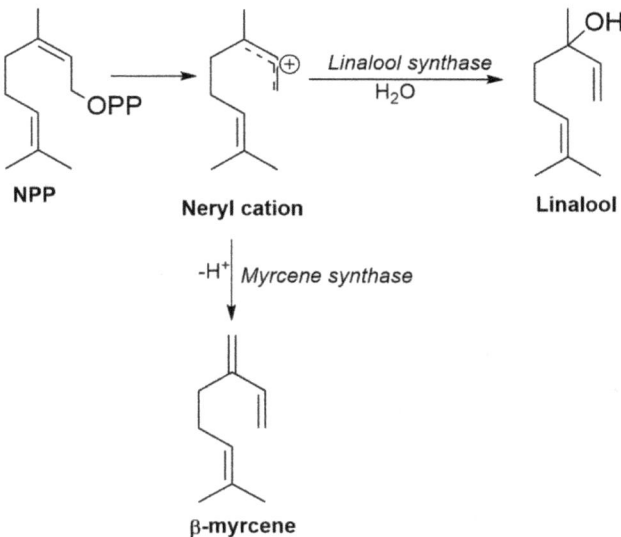

Figure 6.24: Biosynthesis of linalool and *β*-myrcene.

Limonene, *α*-terpineol, cineol, carveol, and carvone are monocyclic monoterpenes that are biosynthesized from the cyclization of NPP that has the favorable Z stereochemistry for cyclization rather than the E stereochemistry in GPP. Limonene is originated via cyclization of NPP by *limonene synthase*. Limonene undergoes allylic hydroxylation by the *hydrolase* enzyme yielding carveol. A *dehydrogenase* enzyme oxidizes the alcohol, carveol, to the ketone, carvone. Cyclization of NPP followed by the attack of water in the presence of *a-terpineol synthase* to afford the tertiary monoterpene alcohol, *α*-terpineol, produces the monoterpene cyclic ether, cineol, by *1,8-cineol synthase enzyme* (Figure 6.25).

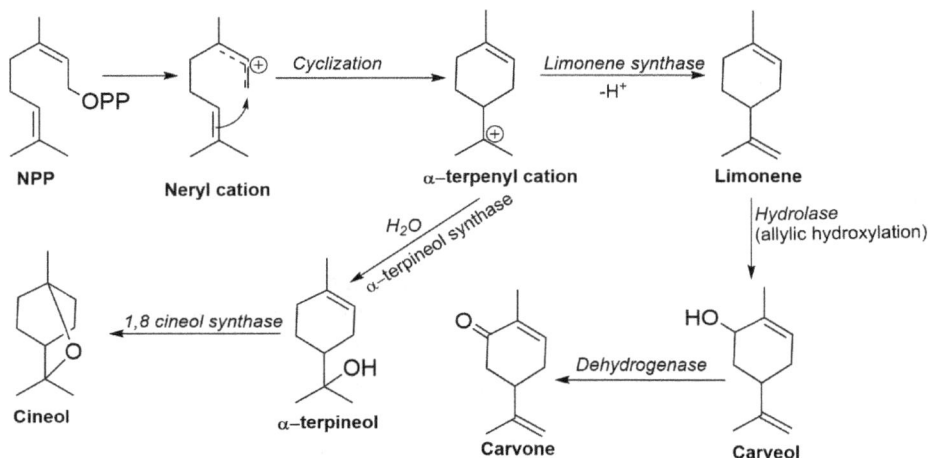

Figure 6.25: Biosynthesis of major monocyclic monoterpenes.

The intermediate *α*-terpenyl carbocation loses a proton and forms a cyclopropane ring by *carene synthase* enzyme and yields the bicyclic monoterpene, Δ³-carene, as shown in Figure 6.26.

a-Pinene and *β*-pinene are two major bicyclic monoterpenes formed by the loss of different protons from the bicyclic *a*-pinyl cation intermediate that originates by the cyclization of neryl cation. *Pinene synthase* is the enzyme responsible for the loss of the protons and the formation of the double bond as shown in Figure 6.27.

The four-membered ring cation, *a*-pinyl cation, intermediate rearranged to the less-strained five-membered ring via 1,2-alkyl shift (Wagner–Meerwein rearrangement) to form fenchyl cation intermediate which transforms to fenchyl alcohol by *fenchyl synthase* enzyme. Oxidation of fenchyl alcohol by *dehydrogenase* enzyme generates the bicyclic monoterpene ketone, fenchone (Figure 6.28).

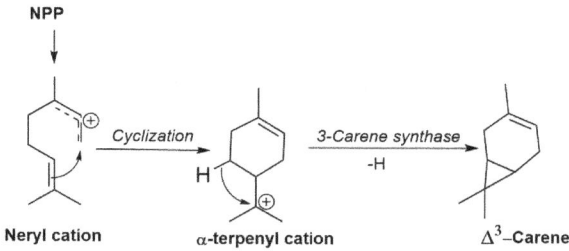

Figure 6.26: Biosynthesis of Δ³-carene.

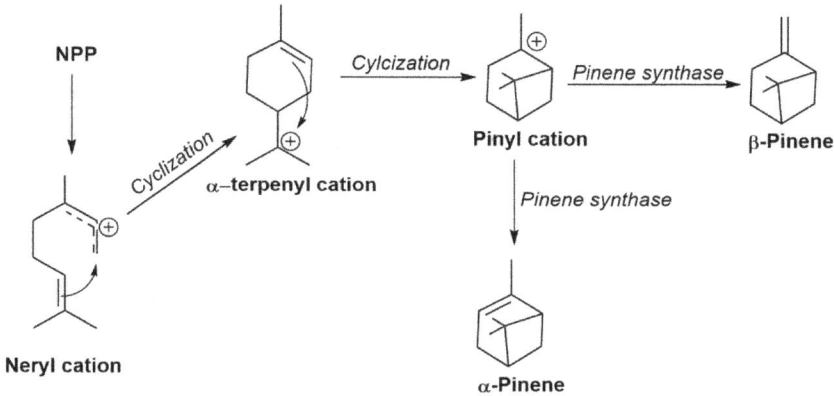

Figure 6.27: Biosynthesis of α-pinene and β-pinene.

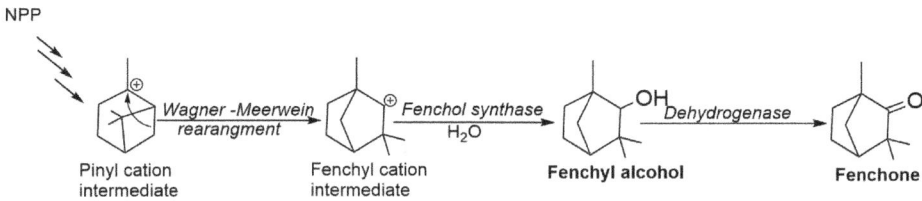

Figure 6.28: Biosynthesis of fenchyl alcohol and fenchone.

6.3.3.2 Biosynthesis of sesquiterpenes

The precursor of sesquiterpenes (C_{15} skeleton) is **farnesyl pyrophosphate or farnesyl diphosphate (FPP)** which is formed by the addition of **IPP** (C_5 unit) to **GPP** (C_{10} unit) by *FPP synthase* (Figure 6.29). More than 45 sesquiterpenes have been reported from cannabis, which can be classified into linear and cyclic sesquiterpenes [1]. *α*-**Humulene**, *β*-**caryophyllene**, and **caryophyllene oxide** are the major cyclic sesquiterpenes present in cannabis. The biosynthesis of these three sesquiterpenes starts with the cyclization of *E,E*-farnesyl cation by the *cyclase enzyme* to form the 11-membered ring humulyl carbocation which yields *α*-humulene by *α-humulene synthase*. Humulyl carbocation may rearrange to form caryophyllene carbocation of nine-membered ring fused to a cyclobutane ring from which *β*-caryophyllene is biosynthesized by the action of *β-caryophyllene synthase*. Caryophyllene oxide arises from *β*-caryophyllene by the epoxidation of the double bond (Figure 6.29).

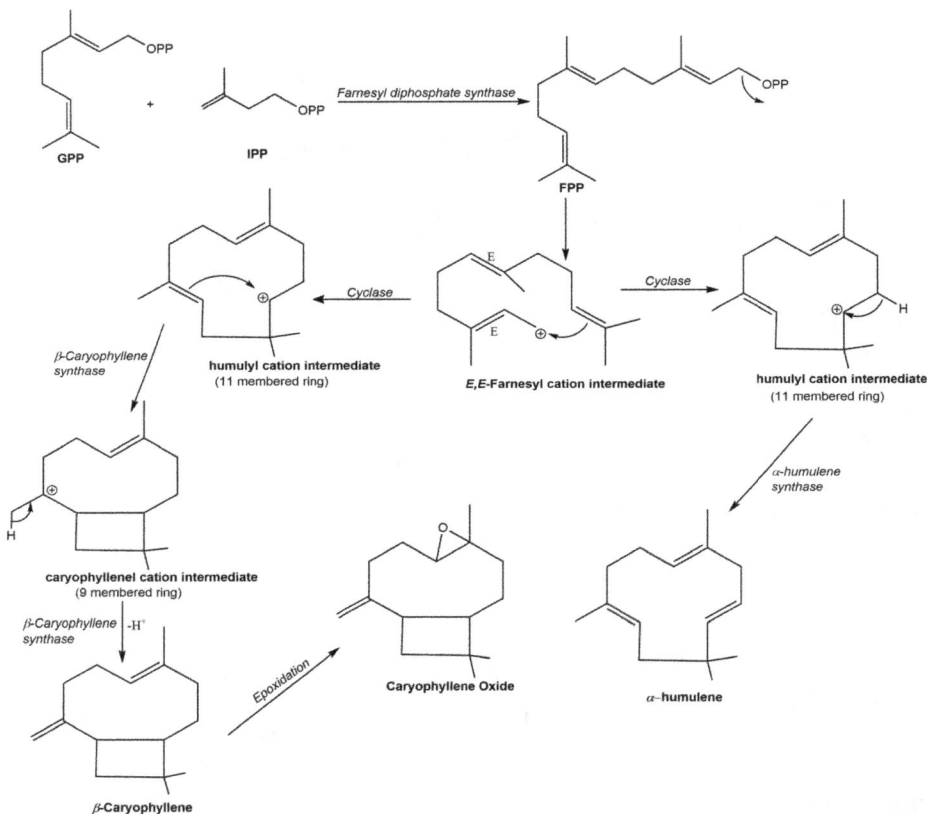

Figure 6.29: Biosynthesis of *α*-humulene, *β*-caryophyllene, and caryophyllene oxide.

6.3.3.3 Biosynthesis of diterpenes

Phytol is an example of diterpenes (C_{20}) isolated from cannabis, and its biosynthesis originates from **geranylgeranyl pyrophosphate (GGPP)** which is formed by the addition of **IPP** (C_5 unit) to **FPP** (C_{15} unit) by *GGPP synthase*. **Phytol** arises as a result of the reduction of **geranylgeraniol** (Figure 6.30).

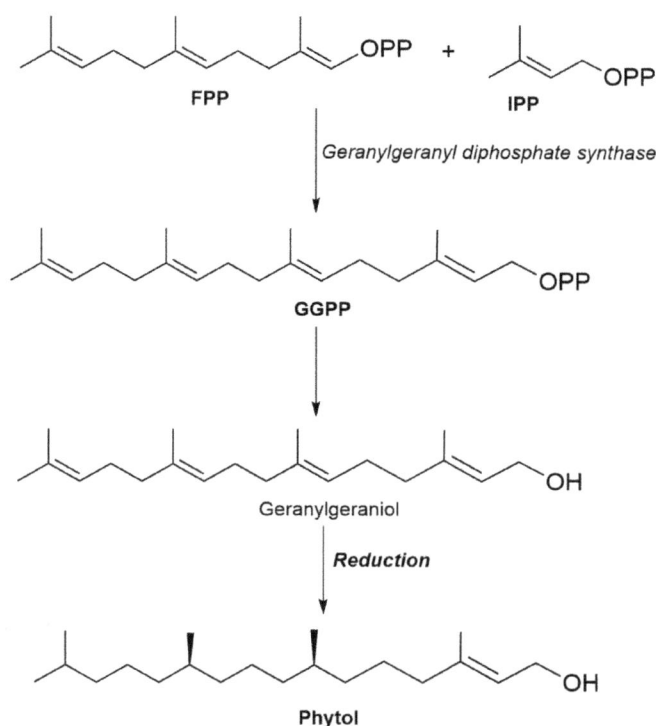

Figure 6.30: Biosynthesis of phytol (diterpene).

6.3.3.4 Biosynthesis of triterpene

Squalene is the precursor of triterpenes (C_{30}) which is formed by the union of two molecules of **FPP** by *squalene synthase enzyme*. **Friedelin** and **epifriedelanol** are two triterpenes isolated from the roots of *C. sativa* [30]. Their biosynthesis starts via epoxidation of **squalene** which forms the intermediate **2,3-oxidosqualene** by *squalene epoxidase*. **Epifriedelanol** is produced from **2,3-oxidosqualene** by enzymatic modifications, including protonation, cyclization, and rearrangement, while its oxidation resulted in the formation of **friedelin** (Figure 6.31).

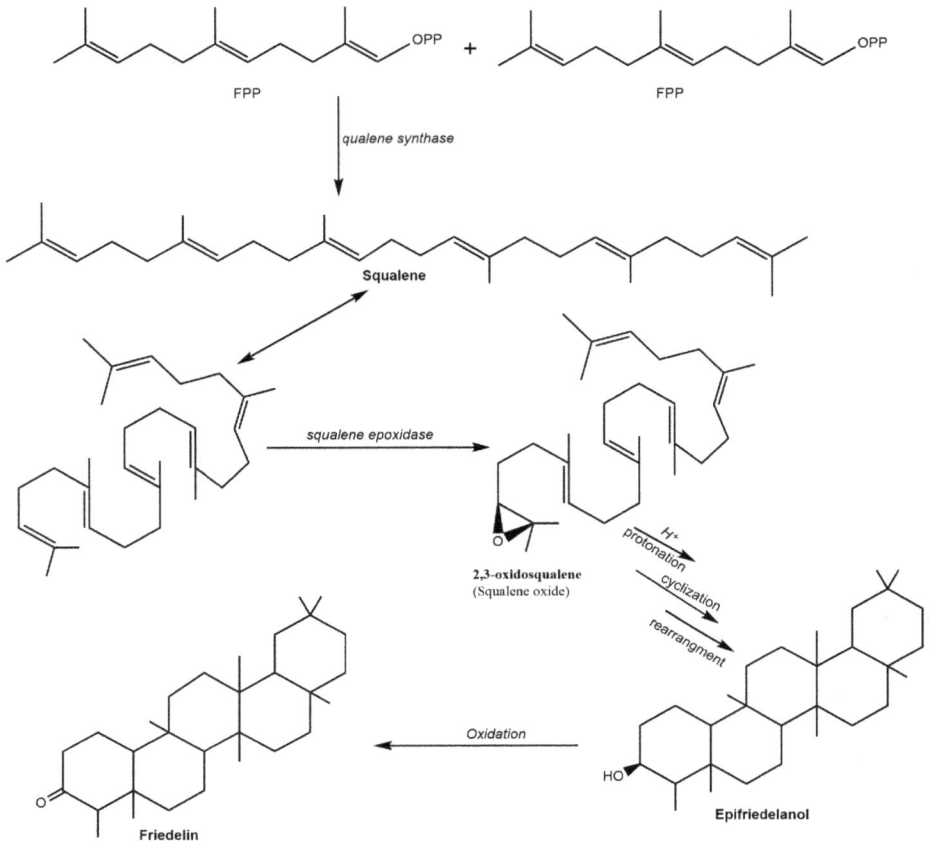

Figure 6.31: Biosynthesis of phytol, friedelin, and epifriedelanol (triterpenes).

6.4 Summary

In this chapter, the biosynthetic pathways leading to the formation of different members of the major chemical classes of the cannabis secondary metabolites have been outlined. The major biosynthetic steps starting with basic primary metabolites, leading to the final secondary metabolites with intermediates in between, and enzymes involved in each transportation step are described for the cannabinoids, non-cannabinoid phenols, flavonoids, and terpenes.

References

[1]   Radwan MM, Chandra S, Gul S, ElSohly MA. Cannabinoids, phenolics, terpenes and alkaloids of cannabis. Molecules 2021; 26: 2774.

[2]   Dewick PM. Medicinal natural products: A biosynthetic approach. Chichester, John Wiley & Sons, 2002.

[3]   Tanaka H, Shoyama Y. Monoclonal antibody against tetrahydrocannabinolic acid distinguishes *Cannabis sativa* samples from different plant species. Forensic Science International 1999; 106: 135–146.

[4]   Andre C, Vercruysse A. Histochemical study of the stalked glandular hairs of the female Cannabis plants, using fast blue salt. Planta Medica 1976; 29: 361–366.

[5]   Stout JM, Boubakir Z, Ambrose SJ, Purves RW, Page JE. The hexanoyl-CoA precursor for cannabinoid biosynthesis is formed by an acyl-activating enzyme in *Cannabis sativa* trichomes. The Plant Journal 2012; 71: 353–365.

[6]   Shoyama Y, Yagi M, Nishioka I, Yamauchi T. Biosynthesis of cannabinoid acids. Phytochemistry 1975; 14: 2189–2192.

[7]   Taura F, Sirikantaramas S, Shoyama Y, Yoshikai K, Shoyama Y, Morimoto S. Cannabidiolic-acid synthase, the chemotype-determining enzyme in the fiber-type *Cannabis sativa*. FEBS Letters 2007; 581: 2929–2934.

[8]   Taura F, Morimoto S, Shoyama Y, Mechoulam R. First direct evidence for the mechanism of. DELTA. 1-tetrahydrocannabinolic acid biosynthesis. Journal of the American Chemical Society 1995; 117: 9766–9767.

[9]   Morimoto S, Komatsu K, Taura F, Shoyama Y. Enzymological evidence for cannabichromenic acid biosynthesis. Journal of Natural Products 1997; 60: 854–857.

[10]   ElSohly MA, Slade D. Chemical constituents of marijuana: The complex mixture of natural cannabinoids. Life Sciences 2005; 78: 539–548.

[11]   Trofin IG, Dabija G, Vaireanu DI, Filipescu L. The influence of long-term storage conditions on the stability of cannabinoids derived from cannabis resin. Revue Chimie Bucharest 2012; 63: 422–427.

[12]   Crombie L, Ponsford R, Shani A, Yagnitinsky B, Mechoulam R. Hashish components. Photochemical production of cannabicyclol from cannabichromene. Tetrahedron Letters 1968; 55: 5771–5772.

[13]   Shani A, Mechoulam R. A new type of cannabinoid. Synthesis of cannabielsoic acid A by a novel photo-oxidative cyclisation. Journal of the Chemical Society D: Chemical Communications 1970; 5: 273–274.

[14]   Küppers F, Lousberg RC, Bercht C, Salemink C, Terlouw J, Heerma W, Laven A. Cannabis – VIII: Pyrolysis of Cannabidiol. Structure elucidation of the main pyrolytic product. Tetrahedron 1973; 29: 2797–2802.

[15]   Yamamoto I, Gohda H, Narimatsu S, Watanabe K, Yoshimura H. Cannabielsoin as a new metabolite of cannabidiol in mammals. Pharmacology, Biochemistry and Behavior 1991; 40: 541–546.

[16]   Yamamoto I, Gohda H, Narimatsu S, Yoshimura H. Mechanism of biological formation of cannabielsoin from cannabidiol in the guinea-pig, mouse, rat and rabbit. Journal of Pharmacobio-Dynamics 1989; 12: 488–494.

[17]   Shoyama Y, Hirano H, Nishioka I. Biosynthesis of propyl cannabinoid acid and its biosynthetic relationship with pentyl and methyl cannabinoid acids. Phytochemistry 1984; 23: 1909–1912.

[18]   Kajima M, Piraux M. The biogenesis of cannabinoids in *Cannabis sativa*. Phytochemistry 1982; 21: 67–69.

[19]   Fritzemeier K-H, Kindle H. 9, 10-Dihydrophenanthrenes as phytoalexins of Orchidaceae: Biosynthetic studies in vitro and in vivo proving the route from l-phenylalanine to dihydro-m-coumaric acid, dihydrostilbene and dihydrophenanthrenes. European Journal of Biochemistry 1983; 133: 545–550.

[20] Crombie L, Mary W, Crombie L. Dihydrostilbenes of Thailand cannabis. Tetrahedron Letters 1978; 19: 4711–4714.

[21] Gehlert R, Kindl H. Induced formation of dihydrophenanthrenes and bibenzyl synthase upon destruction of orchid mycorrhiza. Phytochemistry 1991; 30: 457–460.

[22] Crombie L, Crombie WML. Natural products of Thailand high Δ 1-THC-strain Cannabis. The bibenzyl-spiran-dihydrophenanthrene group: Relations with cannabinoids and canniflavones. Journal of the Chemical Society, Perkin Transactions 1982; 1: 1455–1466.

[23] Bercht C, Van Dongen J, Heerma W, Lousberg RC, Küppers F. Cannabispirone and cannabispirenone, two naturally occurring spiro-compounds. Tetrahedron 1976; 32: 2939–2943.

[24] Nalli Y, Arora P, Riyaz-Ul-Hassan S, Ali A. Chemical investigation of *Cannabis sativa* leading to the discovery of a prenylspirodinone with anti-microbial potential. Tetrahedron Letters 2018; 59: 2470–2472.

[25] Preisigmuller R, Gnau P, Kindl H. The inducible 9, 10-dihydrophenanthrene pathway: Characterization and expression of bibenzyl synthase and S-adenosylhomocysteine hydrolase. Archives of Biochemistry and Biophysics 1995; 317: 201–207.

[26] Cromble L, Mary W, Crombie L, Jamieson SV. Isolation of cannabispiradienone and cannabidihydrophenanthrene. Biosynthetic relationships between the spirans and dihydrostilbenes of Thailand Cannabis. Tetrahedron Letters 1979; 20: 661–664.

[27] Barrett M, Scutt A, Evans F. Cannflavin A and B, prenylated flavones from *Cannabis sativa* L. Experientia 1986; 42: 452–453.

[28] Radwan MM, ElSohly MA, Slade D, Ahmed SA, Wilson L, El-Alfy AT, Khan IA, Ross SA. Non-cannabinoid constituents from a high potency *Cannabis sativa* variety. Phytochemistry 2008; 69: 2627–2633.

[29] Rea KA, Casaretto JA, Al-Abdul-Wahid MS, Sukumaran A, Geddes-mcalister J, Rothstein SJ, Akhtar TA. Biosynthesis of cannflavins A and B from Cannabis sativa L. Phytochemistry 2019; 164: 162–171.

[30] Slatkin DJ, Doorenbos NJ, Harris LS, Masoud AN, Quimby MW, Schiff PL. Chemical constituents of *Cannabis sativa* L. root. Journal of Pharmaceutical Sciences 197; 60: 1891–2.

Amira S. Wanas, Mohamed M. Radwan, Suman Chandra,
Chandrani G. Majumdar and Mahmoud A. ElSohly*

7 Identification and analysis of cannabis

Abstract: The identification of cannabis and the quantitation of the cannabinoids content have always been an important part of cannabis investigations. This has become of even more importance now in light of the legal definition for hemp versus marijuana and the significance of the increase in the tetrahydrocannabinol content and its import on the control level. This chapter elaborates on the different technologies and methods used primarily in the quantitation of the cannabinoids in cannabis products, particularly plant biomass (marijuana and hemp), hashish, and hash oil (cannabis extract). These methods range from the traditional color tests to the currently more sophisticated analytical techniques, namely gas chromatography/flame ionization detector(GC/FID), gas chromatography/mass spectrometry(GC/MS), high-performance liquid chromatography (HPLC), ultra pressure liquid chromatography(UPLC), and high-performance thin layer chromatography(HPTLC).

Keywords: Cannabis identification, microscopy, color test, GC, HPLC, HPTLC

7.1 Introduction

Cannabis is one of the most widely used illicit drugs in the world. It belongs to the family Cannabaceae. Its earliest cultivation for fiber crop is documented in China, wherefrom the crop spread to the Middle East, Europe, and the Americas during the early sixteenth century. Its early medical use is documented by Emperor Shen Neng of China around 12,000 BCE [1].

Cannabis sativa L(*C. sativa*). is one of the oldest medicinal plants and one of the most phytochemically studied. Phytochemical characterization of cannabis highlights the presence of cannabinoids (C21 terpenophenolics), various noncannabinoid constituents including terpenes, flavonoids, spiroindans, dihyrostilbenes, and alkaloids along with 15 other chemical classes totaling more than 550 compounds. The typical C21 terpenophenolic skeleton phytocannabinoids isolated from *C. sativa* represent a class of diverse chemical substances along with their carboxylic acids, analogs, and transformation

*Corresponding author: Mahmoud A. ElSohly, National Center for Natural Products Research and Department of Pharmaceutics and Drug Delivery School of Pharmacy, University of Mississippi, MS 38677, USA, e-mail: melsohly@olemiss.edu, https://orcid.org/0000-0002-0019-2001
Amira S. Wanas, National Center for Natural Products Research, School of Pharmacy, University of Mississippi, University, MS 38677, USA, Department of Pharmacognosy, Faculty of Pharmacy, Minia University, Minia 61519, Egypt
Mohamed M. Radwan, Suman Chandra, Chandrani G. Majumdar, National Center for Natural Products Research, School of Pharmacy, University of Mississippi, University, MS 38677, USA

https://doi.org/10.1515/9783110718362-007

products. The most notable cannabinoid is the tetrahydrocannabinol (Δ^9-THC), the primary psychoactive compound in cannabis. Cannabidiol (CBD) is another major constituent of the plant [2]. Intensive chemical studies have considerably clarified the chemistry of *C. sativa* producing a total of 129 cannabinoids which can be classified into 11 types: Δ^9-THC, $(–)$-Δ^8-*trans*-tetrahydrocannabinol (Δ^8-THC), cannabigerol (CBG), cannabichromene (CBC), CBD, cannabinodiol (CBND), cannabielsoin (CBE), cannabicyclol (CBL), cannabinol (CBN), and cannabitriol (CBT), along with miscellaneous type cannabinoids (Chapter 3). The possession manufacture and distribution of marijuana is an unlawful act by law in the United States according to the Federal Regulations. A strict identification of the marijuana is necessary before proceeding with the persecuting of the case. The microscopical examination of the confiscated botanical parts of marijuana, such as the leaves, seeds, or flowering tops, is one of the methods that help in the identification and authentication of the plant. The presence of cystolith hairs and glandular trichomes is a crucial point of reference in the identification of marijuana leaf, but cannot be used as a sole criterion for marijuana identification. The Duquénois–Levine test is found to be useful in the confirmation of marijuana since none of the 82 species possessing hairs similar to those found on marijuana yield a positive test.

THC is frequently detected in the blood of drivers involved or killed in traffic accidents or suspected of driving while impaired. Therefore, sensitive and accurate methods for determining whether someone is under the influence of cannabis are necessary.

Generally, performing the chemical test is more recommended for the identification of marijuana as it is unrelated to the microscopic appearance and independently confirms the botanical examination. It is more applicable in cases where no recognizable plant material is available as is the case when the evidence is an extract of the plant and its content of cannabinoids, and as in the case of hashish, or in cases where only a residue of resin or a charred ash is available.

7.2 Identification of cannabis

Cannabis plant material can be identified by macroscopical, microscopical, chemical, and analytical methods. Macroscopical and microscopical identifications were discussed in detail in Chapter 2. Herein we will briefly discuss the microscopical identification and focus on the chemical color tests and analytical methods as powerful techniques applied in the identification of cannabis and its products.

7.2.1 Microscopical identification of cannabis

Fresh or dry plant samples can be identified microscopically in the presence of trichomes. Glandular and nonglandular trichromes are identified in different tissues of the plant including leaves, bracts, bracteole, flowers, and stems [3]. Cannabinoids are produced inside the glandular trichomes. The description of the different types of trichrome is discussed in Chapter 2. Although, the presence of cannabis trichomes is essential in the identification of cannabis where plant material is available, it should be confirmed with other methods of identification such as color tests, GC, TLC, or HPLC. Moreover, microscopical examination is not applicable for cannabis products such as extracts, hashish, hash oil, and cannabinoids' isolates.

7.2.2 Color tests

Chemical tests are crucial in the identification of cannabis especially if when coupled with microsocial examination and are also used with plant material and with cannabis products. Duquénois–Levine test, field test, beam test, and Ghamrawy test are examples of these tests which are used for the identification of cannabis plant material and cannabis products. Color tests are rapid and presumptive screening tools used to narrow down the possible identification of an unknown drug sample. While the tests are not specific on their own, they offer the advantage of promptly indicating the presence or absence of a controlled substance.

7.2.2.1 Duquénois–Levine test

This test was developed by Pierre Duquénois in 1930 and was adopted by the League of Nations Sub-committee of Cannabis in 1950 as a specific test for cannabis [4]. In 1960 it was proposed by the United Nations Committee on Narcotics and known Duquénois–Levine test. The original test consists of two steps: the first step involves the addition of an ethanolic solution of vanillin and acetaldehyde to the evaporated petroleum ether extract of the plant or resin material; the second step involves the addition of concentrated hydrochloric acid to the mixture. In the presence of marijuana or marijuana resin, an intense color formation is observed. The modification includes the addition of chloroform which increases the specificity of the test as the phenolic constituents of cannabis give colors which migrate to the chloroform layer [5].

7.2.2.1.1 Samples
Duquénois–Levine test is used for the identification of cannabis plant material, extract, and pure cannabinoids [6].

7.2.2.1.2 Reagents
- Reagent A: 0.5 mL acetaldehyde + 0.4 g vanillin in 20 mL EtOH
- Reagent B: Conc. HCl
- Reagent C: chloroform

7.2.2.1.3 Procedure

1. In a test tube place a small amount (2 mg) of cannabis plant material and shake with 2 mL of reagent A for one min.
2. Add 2 mL of reagent B and shake the mixture and allow standing for 10 min.
3. Add 2 mL of reagent C and mix gently.

7.2.2.1.4 Result

If the lower (chloroform) layer became violet, this indicates the presence of a cannabis product as shown in Figure 7.1.

Rapid Duquénois-Levine Test applied on

High THC variety   High CBD variety

Figure 7.1: Positive Duquénois–Levine test.

7.2.2.1.5 Mechanism of the Duquénois–Levine test

The first step in the mechanism is the protonation of the aldehyde by the acidic solution of hydrochloric acid, which makes the acetaldehyde a stronger electrophile. The protonated acetaldehyde is condensed with vanillin, and the condensation product undergoes further condensation with cannabinoids to form highly conjugated purple color product, so the acetaldehyde is served as a bridge connecting THC with vanillin as shown in Figure 7.2.

Duquénois–Levin test gave positive result with many phenolic compounds containing 1,3-dioxy benzene partial structure (resorcinols) which are found in cannabinoids and flavonoids [7]. Thus this test alone is not specific enough for the identification of cannabis and should be combined with microscopy and/or other chromatographic evidence [8].

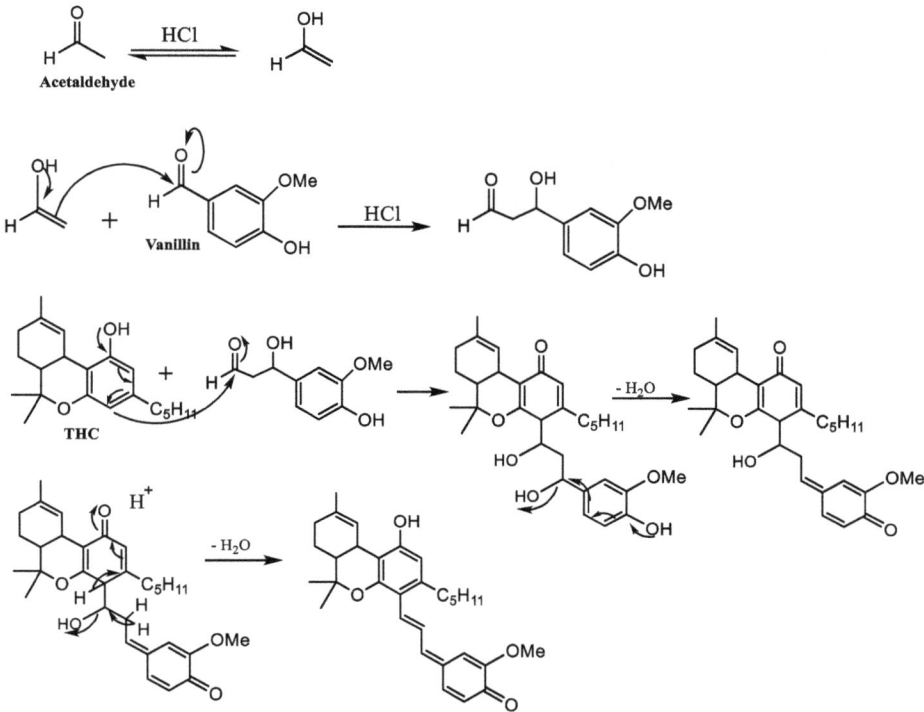

Figure 7.2: Mechanism of Duquénois–Levine test.

7.2.2.2 Beam test

The beam test was introduced in 1911 [9] which is used for the identification of canna-bis plant material and extract.

7.2.2.2.1 Alkaline beam reagent
5% solution of potassium hydroxide in ethyl alcohol (5% KOH in EtOH) [5].

7.2.2.2.2 Procedure
1. Place 1–2 mg of the tested material (powder plant material, resin, or cannabinoid) in a test tube.
2. Add 1–2 mL of petroleum ether and shake and then transfer the extract to an evaporating dish.
3. Evaporate the extract in water bath and add of beam reagent.

7.2.2.2.3 Result

Blue to deep violet color is formed within 30 min. This test gave positive result with CBD and negative with both THC and CBN.

7.2.2.2.4 Mechanism

The color is formed due to the formation of a hydroxyl para quinone and quinone dimer in alkaline medium as shown in Figure 7.3. The presence of both phenolic hydroxyl groups is required to obtain a positive test, so this is used for the identification of CBD and CBG but not THC or CBN [10].

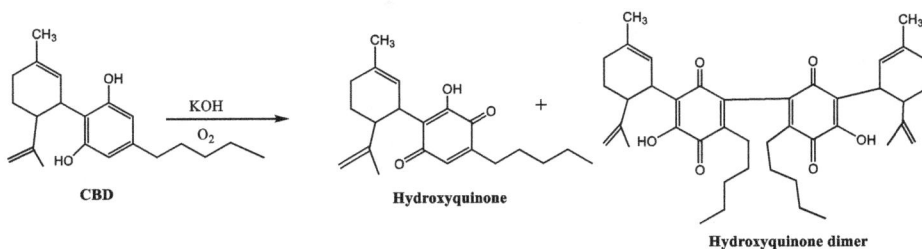

Figure 7.3: Mechanism of beam test.

7.2.2.3 Ghamrawy test

Ghamrawy test was introduced in 1973 and used for the identification of cannabis plant material and hashish.

7.2.2.3.1 Reagent

p-dimethylaminobenzaldehyde (*p*-DMAB) and sulfuric acid (H₂SO₄)

7.2.2.3.2 Procedure

1. Dissolve 1 g of *p*-DMAB in 5 mL H₂SO₄, then add 1 mL of water and allow to cool.
2. Add the above solution to the sample extract (petroleum ether extract of marijuana).
3. Add water to the reaction mixture from step 2.

7.2.2.3.3 Result

A blue-indigo color is formed. This test gave positive test with both CBD and THC.

7.2.2.3.4 Mechanism

Unlike the previous color test, DMAB reacts at the allylic methyl group on the cyclo-hexene ring of Δ^9-THC and CBD. A π conjugation is created with the phenolic ring due to electron rearrangement [11] as shown in Figure 7.4.

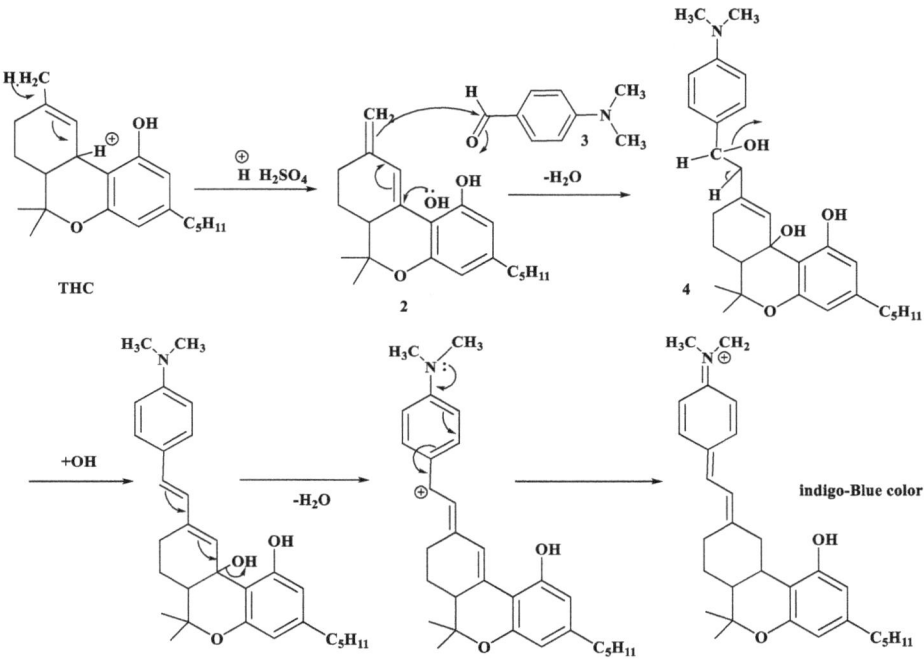

Figure 7.4: Mechanism of Ghamrawy test.

7.2.2.4 Fast blue B test

7.2.2.4.1 Reagent

Fast blue B salt (3,3'-dimethoxybiphenyl-4,4'-bisdiazonium chloride) or fast blue BB salt (4-amino-2,5-diethoxybezanilide diazotated zinc double salt) in basic medium. The two azo dye reagents are designated as FBBS and FBBBS, respectively (Figure 7.5).

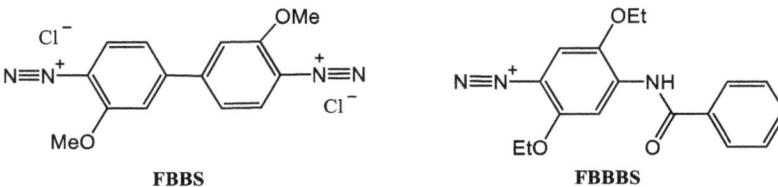

Figure 7.5: Chemical structures of fast blue reagents FBBS and FBBBS.

7.2.2.4.2 Procedure

1. Weight approximately 1 mg of cannabis plant material on an absorbent paper, add two drops of petroleum ether, and wait till it becomes dry.
2. Remove the plant material from the paper and add about 0.1 mg of fast blue reagent (FBBS or FBBBS) to the original spot of the material on the adsorbent paper and then add two drops of water to it.
3. For cannabis extracts or powder or any sticky material, use two thicknesses of absorbent paper and sufficient petroleum ether to moisten the lower paper and perform the test on the lower paper.

7.2.2.4.3 Result

A purple red-colored stain is produced at the original spot on the adsorbent paper (Figure 7.6). Positive test indicates the presence of cannabinoids such as THC, CBN, and CBD.

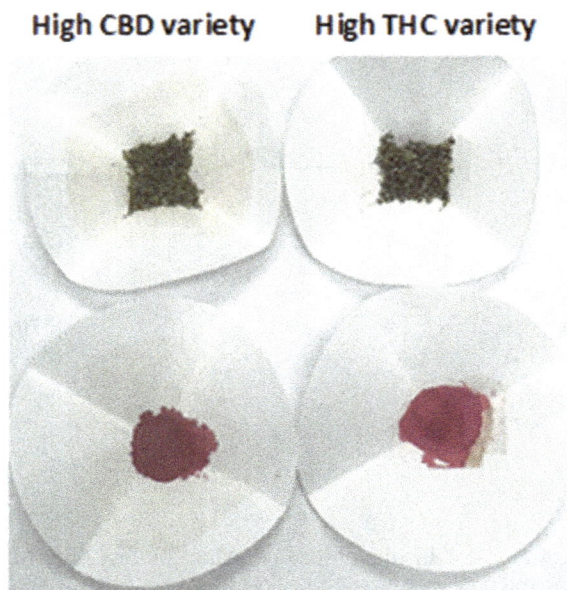

Figure 7.6: Result of FBB color test.

7.2.2.4.4 Mechanism

The mechanism of this color test includes three steps as shown in Figure 7.7. Step 1 is the formation of cannabinoid phenolate ion in the presence of basic medium, which is stabilized by resonance. The second step is the rate determining step which includes the attack of the phenolate electrons to the diazo group of the FBB salt to form a nonaromatic intermediate which undergoes tautomerization to form a color [12].

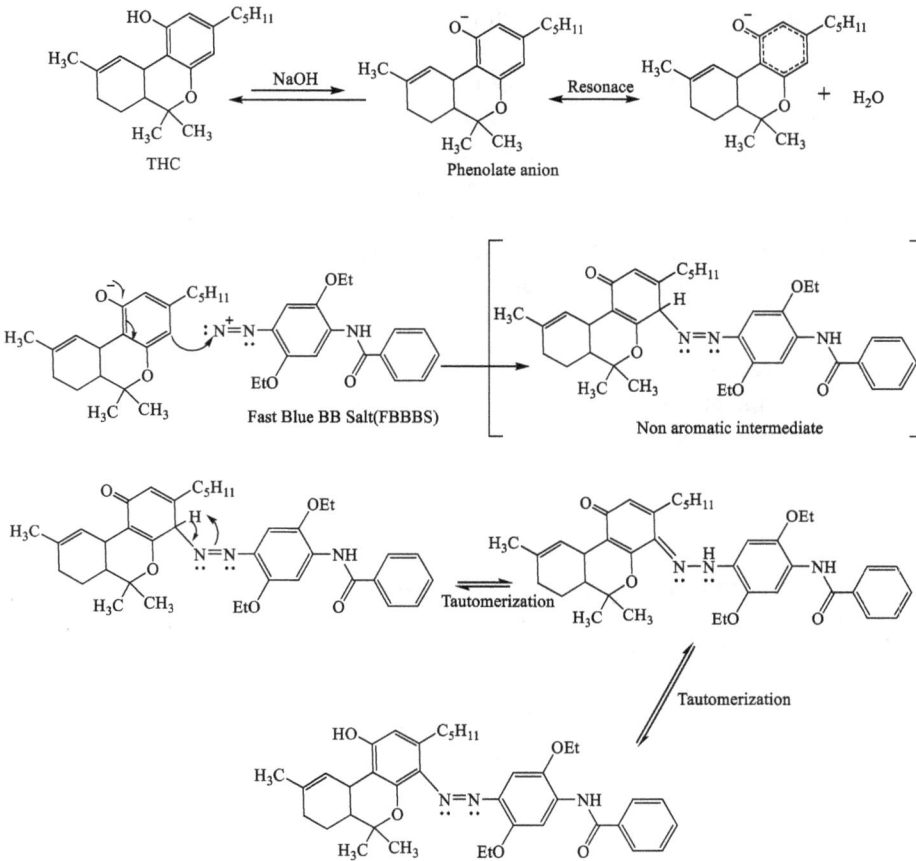

Figure 7.7: Proposed mechanism of the fast-blue color test.

7.2.2.5 4-Aminophenol color test (Swiss test)

The 4-aminophenol (or *p*-aminophenol) color test is used mainly to discriminate between THC-rich and CBD-rich cannabis samples. The test is performed on a test tube or on a spot plate and was originally developed by the Forensic Institute of Zurich (Zurich Police) in Switzerland [13].

7.2.2.5.1 Reagent

Reagent A: 300 mg *p*-aminophenol in 995 mL ethanol and 5 mL 2 N hydrochloric acid.

Reagent B: 30 g sodium hydroxide in 300 mL of water and 700 mL ethanol.

Reagents A and B are stable for six months if they are stored in refrigerator in amber containers.

To execute this test, a small sample of dried or fresh cannabis plant material is transferred into a glass vial and 1–2 mL of solution A (0.3 mg/mL 4-aminophenol in ethanol/isopropanol [95:5; v/v], acidified with 0.5% hydrochloric acid [HCl 2 N]) and four drops of solution B (30 mg/mL sodium hydroxide in ethanol/water [70:30; v/v]) are added. After shaking and leaving to stand for 2 min to allow reaction, the color of the solution is visually evaluated: THC-rich cannabis is indicated by a blue coloration, CBD-rich cannabis by a pink coloration.

7.2.2.5.2 Procedure
1. In a test tube or spot plate, place 5 mg of sample and add reagent A in enough quantity to cover it.
2. Add two to four drops of reagent B to the above mixture and wait for 1–2 min.

7.2.2.5.3 Result
If the concentration of THC in plant material is greater than the concentration of CBD, the color test result is blue and is considered marijuana or THC rich plant. If the concentration of CBD in plant material is greater than the concentration of THC, the color test result is pink and is considered CBD-rich plant or Hemp. The color forms after 2 min.

The decarboxylated CBD and THC plant materials gave more intensive pink and blue colors, respectively, compared to those colors of nondecarboxylated biomasses (Figure 7.8).

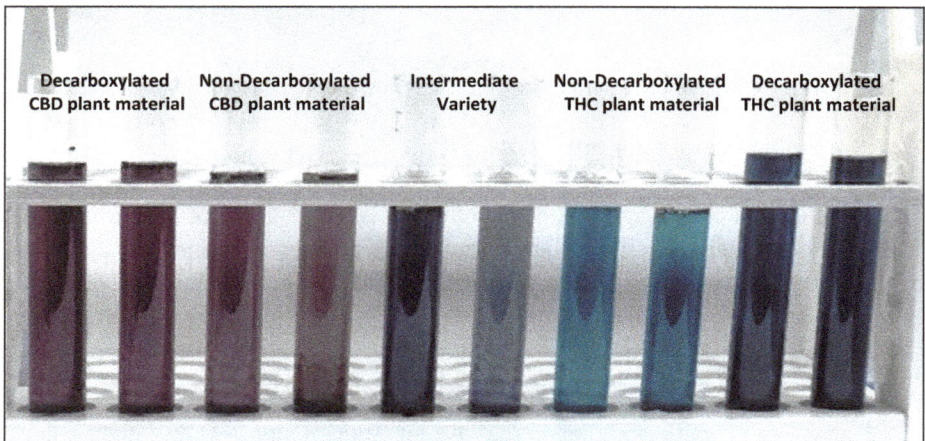

Figure 7.8: Positive 4-aminophenol test.

7.2.2.5.4 Limitation of the test

Plants with equal THC and CBD concentrations (intermediate variety) or a THC/CBD ratio between 0.33 and 3, gave false-negative result as pink within the 2 min time period, and then changed to a mixture of blue and purple after 2 min [14]. In Figure 7.8, the intermediate variety used has THC/CBD ratio of 2:1 and gave a gray color.

7.2.2.5.5 Mechanism

In reagent A, 4-aminophenol is present in a protonated form due to the presence of hydrochloric acid, which is easily deprotonated and oxidized in alkaline medium (reagent B) to quinoneimine intermediate. The intermediate reacts with THC and CBD to generate colored dimers **1** and **2**, respectively, as shown in Figure 7.9.

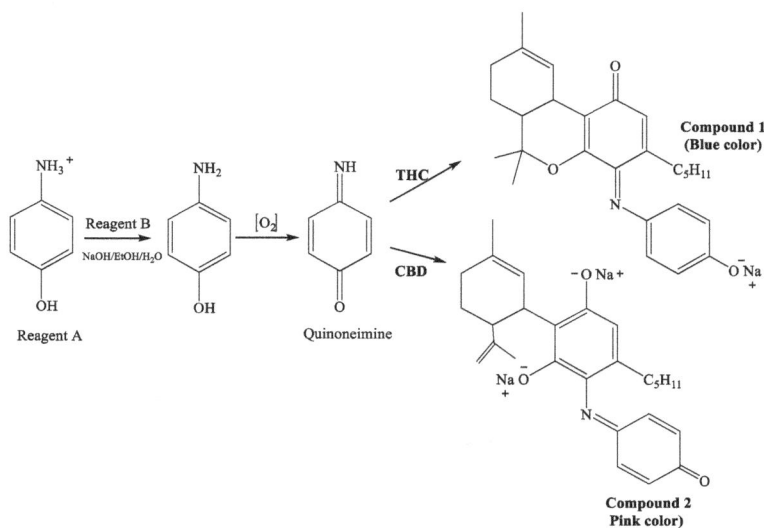

Figure 7.9: Reaction of 4-aminophenol with THC and CBD.

7.3 Chromatographic analysis of cannabinoids in cannabis and cannabis products

The legalization of the medical use of cannabis and its preparations lead to increasing interest in developing analytical procedures needed to ensure the quality in terms of the quantitation of constituents (e.g., cannabinoids and terpenes) and limits of contaminants that might affect the public health. There are three broad categories of cannabis chemotypes based on their chemical profiles: drug chemotype (high THC); intermediate chemotype (THC/CBD); and high CBD chemotype (fiber type or hemp).

Many analytical methods have been developed to determine the chemical profile of cannabinoids in cannabis biomass, extracts, and other cannabis products including GC, HPLC, UPLC, and HPTLC. Those techniques have been widely used with different levels of sensitivity and specificity.

7.3.1 Gas chromatography

GC is the most common analytical technique used for qualitative and quantitative analysis of cannabis chemical components especially cannabinoids and terpenes. In this technique, the separation of these components is based mainly upon differences of the vapor pressure, polarity, and thermal stability in the GC system. GCs are frequently hyphenated to indicate the type of detector used for the analysis. For example, if the detector is a mass spectrometer, the designation would be GC. The advantages of this technique are being fast, simple, and sensitive for the determination of the total cannabinoids (neutral and acidic) and terpenes. The decarboxylation of the acidic cannabinoids due to the high injector temperature does not permit their determination unless chemical derivatization is performed, such as silylation prior to injection into the GC.

7.3.1.1 Methodology

The sample is first introduced into the GC with a syringe or transferred from an autosampler into the GC inlet through a septum, which is connected to the analytical column, which contains the stationary phase coated on the inside walls. The analytical column is held in the column oven which is heated during the analysis to elute the volatile components. The outlet of the column is inserted into the detector which responds to the chemical components eluting from the column to produce a signal. The signal is recorded by the acquisition software on a computer to produce a chromatogram. There are many types of GC detectors, for example: those that respond to C–H bonds like the FID; those that respond to specific elements, for example, nitrogen or phosphorus (NP detector); and those that respond to specific properties of the molecule, like the ability to capture an electron, as is used with the electron capture detector and finally an MS. The main components of GC chromatograph are illustrated in Figure 7.10.

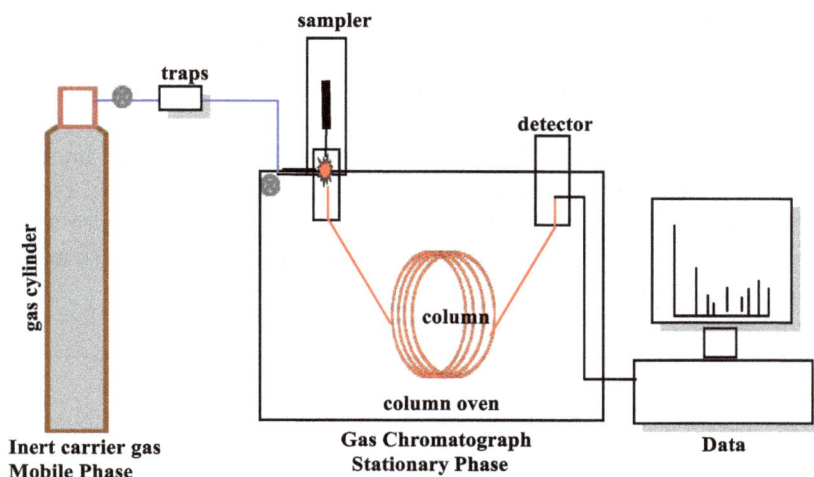

Figure 7.10: A simplified diagram of a gas chromatograph.

7.3.1.2 Applications

7.3.1.2.1 GC analysis of neutral cannabinoids

Different GC–FID and GC/MS methods have been reported for identification and quantification of cannabinoids and terpenes in cannabis. The three major cannabinoids (Δ^9-THC, CBD, and CBN) in some hemp food products including beer, pastilles, liqueur, seeds, and oil were quantitatively analyzed by GC/MS using fused silica capillary column (HP-5 MS, 30 m × 0.25 mm i.d, film thickness 0.25 μm). Oven temperature was adjusted at 120 °C for 2 min, then increased to 290 °C at 20 °C/min and held for 10 min. The electron-impact mass spectra of the analytes were recorded in the total ion monitoring mode (scan range 40–550 M), and Δ^8-THC was used as an internal standard [15].

GC/FID method was used for the quantification of the cannabinoids content in a Japanese Cannabis plant material for Δ^9-THC, CBD, CBC, and CBN. The run was done on an HP-5 MS column (30 m × 0.25 mm × 0.25 μm) using 5α-cholestane as an internal standard. Oven temperature was programmed from 50 °C to 250 °C, and injection volume was 1 μL. The time of the run was 30 min [16].

A validated GC/FID method of quantification of Δ^9-THC, CBD, and CBN in 52 seized marijuana samples and two hashish samples in Brazil was achieved using an HP-5 fused-silica GC column (30 m × 0.32 mm i.d., 0.25 μm film thicknesses, Agilent). The temperature of the injection port and detector was 270 °C and 280 °C, respectively. The oven temperature was maintained at 150 °C for 1 min; programmed at 15 °C/min to 250 °C followed by a hold time of 13 min and using diazepam as internal standard. This rapid and simple method was able to differentiate between different cannabis phenotypes [17].

Illicit cannabis products including sinsemilla, ditchweed, Thai sticks, hashish, and its oil seized by the U.S. Drug Enforcement Administration (DEA) were quantitatively analyzed using fast, sensitive, precise, and accurate GC/FID method for its seven cannabinoids content (Δ^9-THC, Δ^8-THC, CBD, CBN, CBG, CBC, and THCV). The used validated method was carried on DB-1 MS columns (15 m × 0.25 mm × 0.25 µm), injector temperature, 240 °C; detector temperature, 270 °C; oven program, 170 °C (hold 1 min) to 250 °C at 10 °C/min (hold 3 min); programmed to run for 12 min, using 4-androstene-3,17-dione as the internal standard [18]. The GC chromatogram of those seven cannabinoids is shown in Figure 7.11.

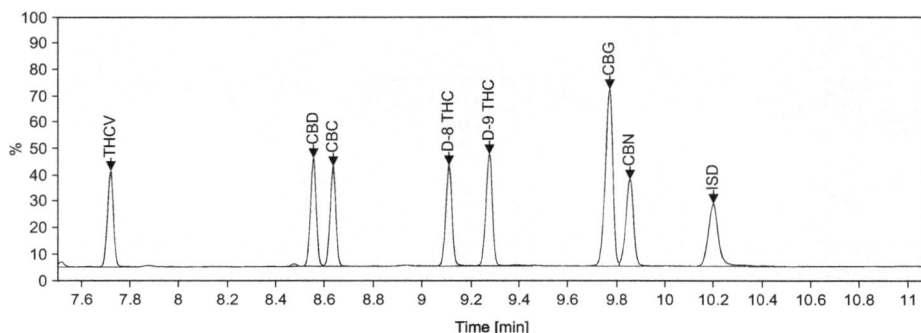

Figure 7.11: GC chromatogram of seven major cannabinoids.

7.3.1.2.2 GC analysis of cannabinoid acids

The GC technique is not the proper technique for the analysis of cannabinoid acids, as the high temperature of the injection port causes decarboxylation of the acids. Therefore, a new validated GC/FID method was developed for the analysis of both acids and neutral cannabinoids in the same sample. This technique depends on the derivatization of the cannabis samples with trimethylsilyl groups that are incorporated at the phenolic hydroxyl and carboxyl groups so that the cannabinoid compounds have sufficient volatility to be easily detected on GC–FID. Silylation was carried out by reaction with *N,O*-bis(trimethylsilyl)trifluoroacetamide. The method is accurate, reproducible, and rapid. This validated method was applied to identify and quantify 13 cannabinoids including four cannabinoid acids (CBDVA, THCAA, CBDA, and CBGA), along with nine neutral cannabinoids (Δ^9-THC, Δ^8-THC, CBN, CBG, CBL, CBDV, THCV, CBC, and CBD) in different types of cannabis plant materials. The oven time program began at 190 °C for 1 min before ramping at a rate of 30 °C/min to 230 °C. The oven was kept at 230 °C for 2 min before ramping at a rate of 5 °C/min until reaching 250 °C. After holding for 1 min, the oven temperature increased at 20 °C/min to 300 °C where it was held for 2.75 min and then recycled back down to 190 °C. The total run time was approximately 17.5 min. The detector temperature was 300 °C, and the hydrogen, air, and make-up gas flow rates were 40, 500, and 27 mL/min, respectively. 4-androstene-3, 17-dione was used as internal standard [19]. The GC chromatogram of the silylated cannabinoids (cannabnoids–TMS derivatives) is shown in Figure 7.12.

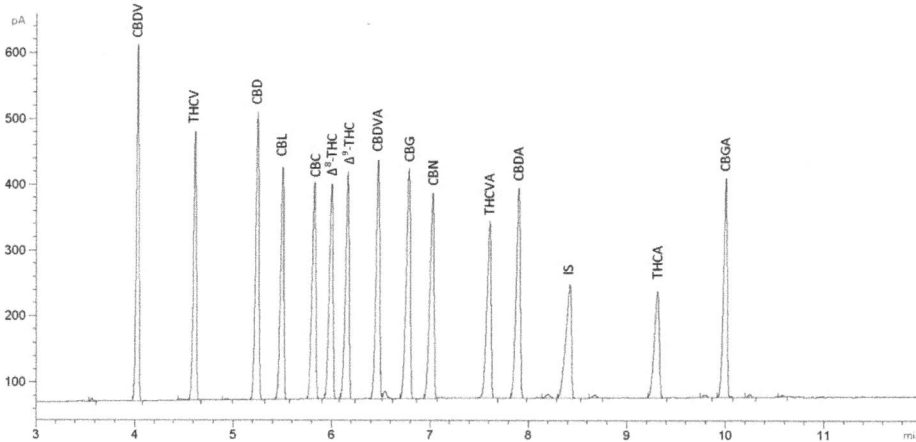

Figure 7.12: GC–FID chromatogram of cannabinoids–TMS.

7.3.1.2.3 GC analysis of terpenes

Terpenes are responsible for the aroma of cannabis. Along with the cannabinoids, determination of the terpenes in cannabis is important in choosing the best *C. sativa* strain. A GC–MS method was developed for the quantification of the 10 most prevalent terpenes in cannabis plant material namely α-pinene, β-pinene, β-myrcene, limonene, terpinolene, linalool, α-terpineol β-caryophyllene, α-humulene, and caryophyllene oxide. Samples were prepared by extraction of the plant material with ethyl acetate containing *n*-tridecane (C13 hydrocarbon) solution (100 µg/mL) as the internal standard. *n*-tridecane was selected as the internal standard since it was found experimentally that its retention time falls between the mono (C10) and sesquiterpenes (C15) and the fact that it was not present in *C. sativa* plant extracts [20].

DB-5 MS capillary column (30 m × 0.25 mm I.D., 0.25 µm film thickness) was used. Helium was used as the carrier gas at a flow rate of 1 mL/min. The inlet temperature was 250 °C with a split ratio of 15:1. The injection volume was 2 µL. The oven temperature program started at 50 °C (held for 2 min), then ramped up to 85 °C at a rate of 2 °C/min, and to 165 °C at 3 °C/min. The postrun temperature was 280 °C for 10 min. The mass spectrometer was set in full scan mode from 40 to 450 amu. The ionization energy was 70 eV. The ion source temperature was 230 °C and the quadrupole temperature was 150 °C and the transfer line temperature was 280 °C. The solvent delay was set to 4 min. Figure 7.13 shows the GC chromatogram of the analyzed terpenes and *n*-tridecane (internal standard).

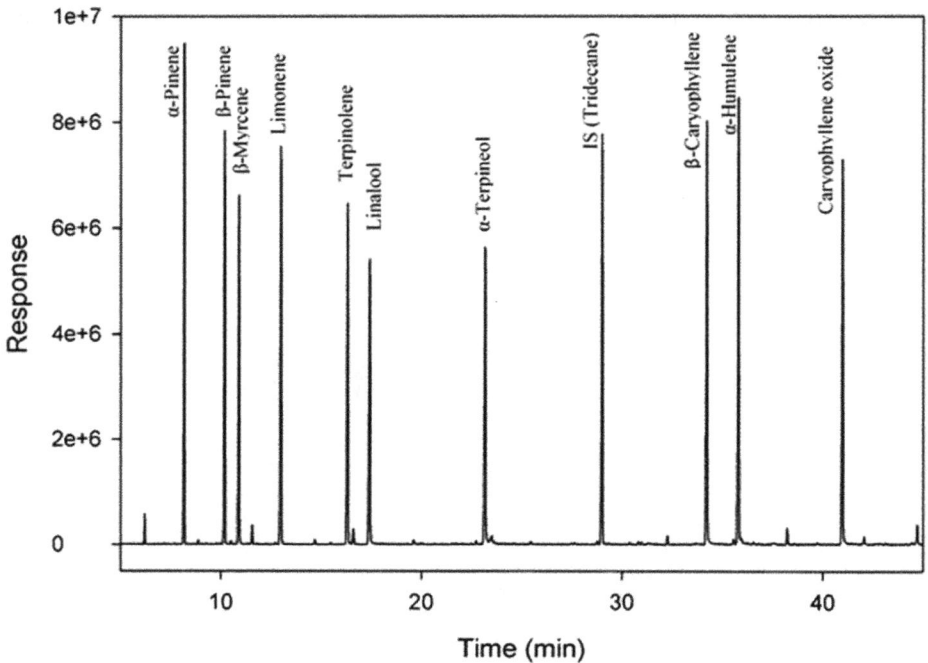

Figure 7.13: GC chromatogram of terpenes.

7.3.2 High-performance liquid chromatography

HPLC is one of the analytical techniques in which the chemical structure of the cannabinoids remains intact (since no heat is applied), which permits analysis of both neutral and acidic cannabinoid in the same run. It has the disadvantage of lack of insufficient resolution for the separation of the whole array of cannabinoids and other noncannabinoids present in the plant material extracts.

7.3.2.1 The apparatus of the HPLC

The HPLC system consists of a solvent delivery pump that delivers the mobile phase at an adjusted flow rate, a sample injector, a column oven, a detector, and a data processor (Figure 7.14).

In order to avoid flow rate fluctuations and baseline noise/drift, a solvent degassing is used to remove air bubbles from the mobile phase, helium in the mobile phase at a very low rate. The column is placed in a column oven to keep the temperature constant, as temperature fluctuations can affect the separation of compounds on the column.

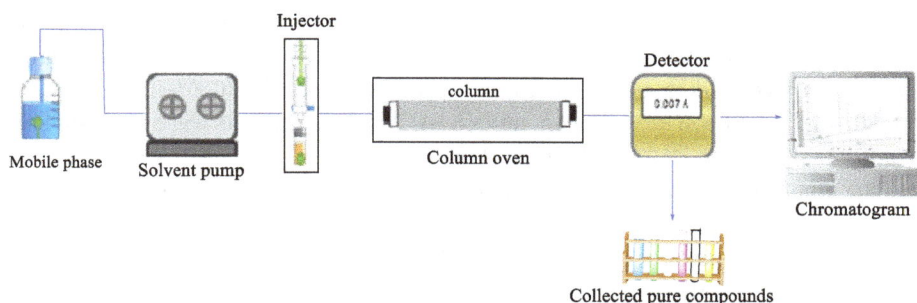

Figure 7.14: High-performance liquid chromatography (HPLC) system.

Compounds eluted from the column are detected by a detector which is placed downstream of the column. A workstation processes the signals from the detector to obtain a chromatogram to identify and quantify the compounds. Since sample compounds characteristics can be very different, several types of detectors have been developed such as UV-absorbance detector, fluorescence detector, and evaporative-light-scattering detector (ELSD). Multiple detectors in series are sometimes used, for example, and a UV and/or ELSD detector may be used in combination with an MS to give more comprehensive information about an analyte. The application of combining a mass spectrometer to an HPLC system is called LC/MS.

Each HPLC detector has its own advantage and limitation. UV detectors require that the analyte must possess a chromophore to be detected. Fortunately, all cannabinoids and their acids have aromatic rings and therefore UV absorbance. Terpenes on the other hand do not. Compounds that lack chromophore could be analyzed using evaporative light scattering (ELS) detector. This is a universal detector in which the molecular mass of the compounds eluting off the column into the mobile phase. Finally, mass spectrometric detection offers the advantage of specificity as to the identity of the elating compounds and is applicable whether the analyte has a chromophore or not.

7.3.2.2 Applications

A validated HPLC method was used for the analysis of THCA, CBDA, THC, CBD, CBG, CBC, Δ^8-THC, and CBN in two cultivars of Cannabis from California. Separation was achieved on a Poroshell 120 EC-C18 column (2.7 μm, 150 × 2.1 mm id, using photodiode array detector (PDA), and 214 nm for quantification). Gradient eluent consisted of 0.1% formic acid in water (solvent A) and 0.1% formic acids in acetonitrile (solvent B) was used. Ibuprofen was used as internal standards. Selectivity, linearity, accuracy (recovery and percentage relative bias), and repeatability precision (RSDr) of the method were determined. This method provided baseline resolution of the eight cannabinoids in 17 min [21].

Quantification of eleven cannabinoids, in three different varieties of cannabis as well as in seizures made by the DEA was performed at the University of Mississippi using a validated HPLC method. The cannabinoids included Δ^9-THC, Δ^8-THC, CBG, CBC, CBD, CBDA, CBL, CBN, THCV, THCAA, and CBGA. The cannabinoids were separated on a Luna C18 (2) column (150 × 4.60 mm id, 3 μm particle size).

The mobile phase consisted of (A) 0.1% (v/v) formic acid in water and (B) 0.1% (v/v) formic acid in acetonitrile with gradient elution program. UV spectra were recorded from 210 to 400 nm, and the quantification wavelength was set at 220 nm. Run time was 22.2 min and the method was described as accurate, fast, and reliable and could be used for routine analysis of cannabis (Figure 7.15) [22].

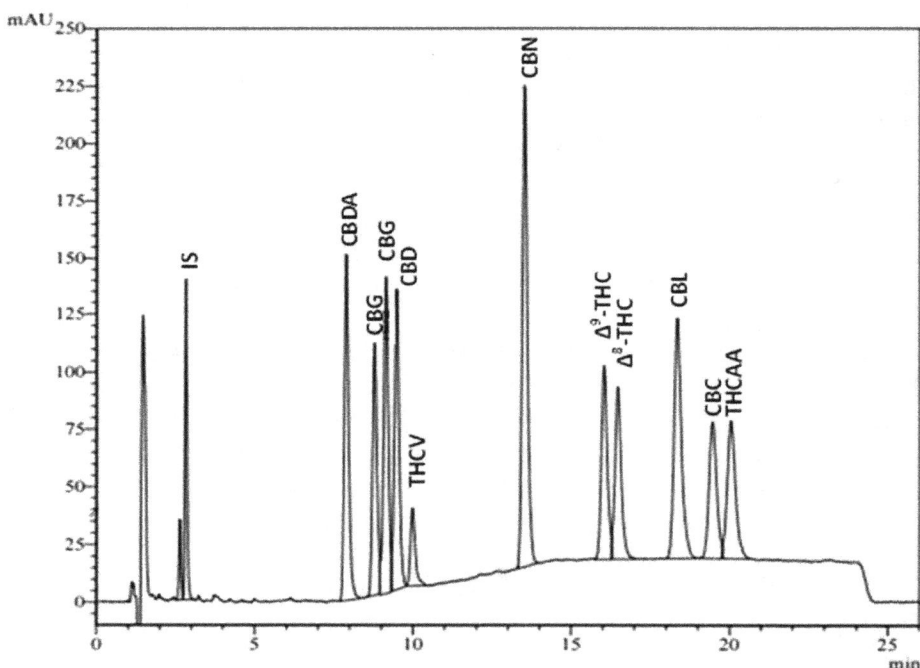

Figure 7.15: HPLC chromatogram of 11 cannabinoids.

Wang et al. [23] determined the concentrations of three cannabinoids, Δ^9-THC, CBD, and CBN in 13 Cannabis edible and beverage samples. The samples include baked goods, chocolate bars, and hard candies. LC–MS/MS was used in the positive electrospray ionization mode (ESI–MS) and C18 HPLC column (100 mm × 2.1 mm × 3 μm particle size) column. The mobile phase consists of 10 mM ammonium acetate and methanol with 0.1% formic acid in a gradient manner. The method was described as QuEChERS (quick, easy, cheap, effective, rugged, and safe).

A manual prepared by the United Nations, Division of Narcotic Drugs has described an HPLC method for the analysis of CBD, CBN, THC, and THCA in homogenous herbal cannabis using 250 mm × 4.0 mm RP-8 (5 µm) column and an isocratic mobile phase [acetonitrile: water (8:2 v/v)]. The total run time was 8 min. The quantitation was carried out at two wave lengths of 220 and 240 nm [24].

A validated HPLC/DAD method developed by De Backer et al. [25] has been chosen by the American Herbal Pharmacopeia for qualitative and quantitative determination of Δ^9-THC. THCA, CBD, CBDA, CBG, CBGA, CBN, and Δ^8-THC in eight cannabis samples of drug type. HPLC column C18 (4.6 mm × 150 mm × 3.5 µm particle size) and a gradient mobile phase composed of 50 mM ammonium formate (pH 3.75) and acetonitrile were used. Diazepam was the internal standard. Neutral cannabinoids were detected at 228 nm while acidic cannabinoids at 270 nm. It is an accurate method for not only the quantification of major cannabinoids in cannabis plant but also it can be used for the determination of plant phenotype. The disadvantage of this method is relatively long run (36 min) [25].

The United States Pharmacopeia has reported specifications necessary to define cannabis quality attributes. HPLC was one of the chromatographic methods used to verify the identity of the chemotype based on the presence and the percent of the THC and CBD. Standard solutions of THC, CBD, and cannabinoid acids mixture were used as standards. Sample solution was prepared using cannabis inflorescence in methanol, homogenized in a high-throughput homogenizer for 1 min at 1,500 rpm. The extract was then filtered through submicron pore filter to be injected for analysis. The HPLC conditions are: detector: UV with diode array detection, 222 nm, column: 4.6 mm × 15 cm, 2.7 µm C18, hard core with superficially porous shell, L1, column temperature: 40° and flow rate: 1.5 mL/min injection volume: 5 µL. Thirteen cannabinoids (acids and neutrals) have been separated (Figure 7.16) [3].

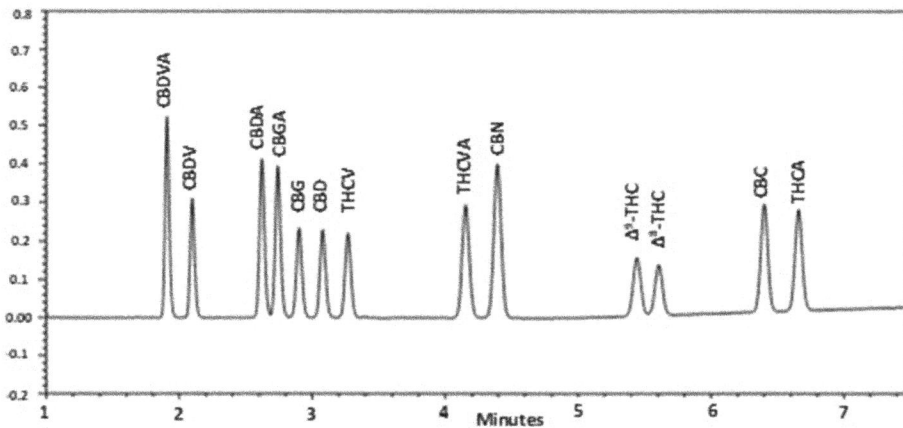

Figure 7.16: HPLC profile of cannabinoids.

7.3.2.3 Analysis of flavonoids

An HPLC method with UV/DAD, ESI–MS, and MS/MS detection was developed and validated for the analysis of hemp-specific flavonoids, combined with a selective and simple extraction procedure using six samples of fiber-type hemp female inflorescences. The HPLC method applied for profiling of the content of cannflavins A and B. Cannflavin A was observed to be the main compound in almost all the samples. The developed HPLC method can be applied for ensuring the quality of both hemp plant material and related pharmaceutical products.

A two-step extraction of the plant material with *n*-hexane followed by extraction with acetone was selected as the optimum protocol for flavonoids extraction to reduce any interference from cannabinoids in the hemp extract.

The applied HPLC method was optimized to get a good resolution of peaks belonging to cannflavins A and B using gradient elution of acidic mobile phase that provides good separation, enhanced resolution, better peak shape, and improvement of their ionization in HPLC–ESI–MS and MS/MS experiments. The chromatographic analysis was carried out on an Ascentis Express C18 column (150 mm × 3.0 mm I.D., 2.7 μm), with a mobile phase composed of 0.1% HCOOH in both (A) H_2O and (B) ACN. The gradient elution was modified as follows: 0–5 min 40% B, 5–20 min from 40 to 80% B, and 20–35 min from 80% to 90% B, which was held for 10 min. The postrunning time was 10 min. The flow rate and the sample injection volume were kept at 0.4 mL/min and 3 μL, respectively, and column was kept at room temperature.

The quantification of cannflavins A and B of the tested six hemp samples was carried out by combining the information obtained from UV/DAD and MS detectors. Cannflavins A and B (Figure 7.17) showed the first λ_{max} at 213–215 nm, a second one at around 273–274 nm, and finally, one at 342 nm. The amount of cannflavin A in the tested samples raised from 0.006% to 0.032% while the content of cannflavin B was found to be in the range of 0.003% to 0.041% [26]. The measured amounts of the cannflavins were in good agreement with the published data by Izzo et al. [27].

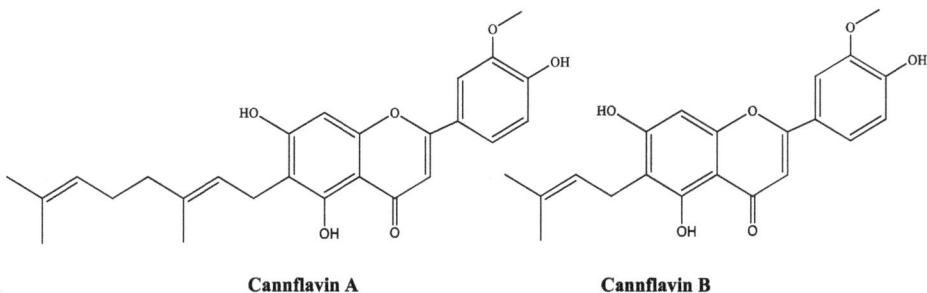

Cannflavin A   **Cannflavin B**

Figure 7.17: The chemical structures of cannflavins A and B.

7.3.3 Ultraperformance liquid chromatography

UPLC offered the advantages of increased sensitivity and resolution together with reduced analysis time by using columns with particle size of 2 μm and smaller. Thus, a
greater resolution is achieved between peaks, or the same resolution can be achieved
in less time. Due to the better sensitivity enhanced resolution, UPLC was widely used
in forensic chemistry to quantitate cannabinoids and their metabolites in biological
fluids (urine, blood, and saliva). Few publications on the application of UPLC for the
analysis of cannabis products were reprinted in the literature.

7.3.3.1 Methodology

The analytical UPLC instrument consists of the same parts as HPLC instrument. The
main differences between UPLC and HPLC are using short columns packed with
smaller particles (sub-2 μm), the much-increased pressure limit on the solvent pump,
and the decreased use of solvents compared with conventional HPLC. Moreover, the
UPLC system enables the detection of analytes at very low concentrations because of
the improved signal-to-noise ratio. The injection volume in UPLC can be significantly
reduced without loss of sensitivity [28].

7.3.3.2 Applications

UPLC/UV and UPLC–MS–MS were described and validated by Seok et al [29]. for the analysis of cannabinoids in different types of food and as well in dietary supplements (tablets, capsules, powders, liquids, cookies, and candies). The UPLC/UV validation method
was performed on an Acquity UPLC/UV system using Waters Acquity UPLC HSS C18 column (2.1 mm × 150 mm, 1.8 μm particle size) with a flow rate of 0.18 mL/min, and the UV
detection was set at 210 nm. The mobile phase was gradient and consisted of 25 mM sodium phosphate and 0.01% sodium hexane sulfonate in deionized water adjusted to pH
3 with phosphoric acid (solvent A) and acetonitrile (solvent B). LC–MS–MS analysis was
carried out on a Waters Acquity UPLC BEH C18 column (2.0 mm × 100 mm, 1.7 μm) with
a flow rate of 0.25 mL/min. The mobile phase was composed of solvent A (0.1% formic
acid in distilled water) and solvent B (0.1% formic acid in acetonitrile). MS was conducted in ESI mode. Both methods were validated for linearity, precision, and accuracy.
This method can be used for rapid and accurate screening of cannabinoids present in
food products.

A simple, fast, and efficient method was developed for the analysis of 30 cannabis
plant samples (flowering buds, hashish, and leaves) at the University of Mississippi
using ultra high-performance supercritical fluid chromatography coupled with PDA
and ESI/mass spectrometry (ESI–MS) detection. Nine cannabinoids including CBD,

Δ^8-THC, THCV, Δ^9-THC, CBN, CBG, THCA-A, CBDA, and CBGA were quantitatively determined. The chromatographic separation was achieved using a Waters ACQUITY UPC^2 BEH 2-EP (2-ethylpyridine) column with dimensions of 150 × 3.0 mm i.d. and 1.7 μm particle size. The mobile phase consisted of CO_2 as solvent A and isopropanol: acetonitrile (80: 20) with 1% water as solvent B. The PDA was set to scan from 190–400 nm, and 220 nm was used for the quantification. Mass spectrometry was performed using a Waters ACQUITY single quadrupole mass spectrometer. The MS ESI source was operated in full scan mode (positive and negative) in a mass range from 100 to 800 amu. The validated method is more sensitive, and the run time is shorter compared to GC/MS methods. Multivariate statistical analysis including principal component analysis and partial least squares-discriminant analysis was used to differentiate between the cannabis samples [30].

CBD, CBDA, CBG, CBGA, THC, and THCA were detected in in dietary supplements, food, and beverages using HPLC–ESI–MS/MS in negative ion mode on a linear ion trap quadrupole (QTRAP) mass analyzer using Phenomenex Kinetex EVO C18 column (100 mm × 2.1 mm; particle size: 5 μm; Phenomenex) [31].

7.3.4 High Performance Thin Layer Chromatography

HPTLC is an applied technique for the analysis of herbal drugs which allows analysis of many samples in parallel with the possibility of multiple detections. It is a sophisticated form of thin layer chromatography (TLC) with superior and advanced separation efficiency and detection limits and is often an exceptional alternative to HPLC and GC. HPTLC is also known as flat-bed chromatography or as planar chromatography with many applications in phytochemical analysis, herbal drug quantification, and finger print analysis. Normal and reversed-phase HPTLC plates could be used.

The HPTLC works on the principles of adsorption, where the analytes move according to their affinities toward the stationary phase and the mobile phase as the solvent flows through the stationary phase by capillary action. Component with higher affinity toward the stationary phase travels slower through the stationary phase, while low-affinity components travel rapidly with the mobile phase resulting in physical separation of the components of the mixture to be analyzed.

The main advantage of the HPTLC technique is its capability to analyze multicomponents of multiple samples simultaneously, on the same plate. The lower prices of the precoated plates make it a more economic method. HPTLC also allows for two-dimensional separations at no risk of contamination due to the use of the freshly prepared mobile phase and stationary phase. Moreover, neither filtration nor degassing of mobile phases is required as in case of HPLC. The method is highly sensitive, reproducible, and precise as compared to conventional TLC and minimized exposure risk of toxic organic effluents and reduction of disposal problems and environmental pollution [32]. Short separation bed is the major disadvantage of HPTLC in addition to a limited number of samples per plate can be tested.

7.3.4.1 Methodology

HPTLC is chromatographic technique based on the principle of planar chromatography but more advanced than TLC with better resolution and reproducibility. HPTLC precoated glass plates with silica gel 60 F254S or reversed phase are used. Samples are applied automatically using a sampler applicator as 6 mm band. The run of the samples vertically is done in a chromatographic development chamber. The detection of the tested compounds is carried out in the visualizer and then in the scanner with UV detection. The software organizes the workflow of the HPTLC analysis and manages data evaluation (Figure 7.18).

Applicator   Developer   Visualizer   Scanner   MS Interface   Derivatizer

Figure 7.18: A schematic diagram for the components of CAMAG®-HPTLC instrument.

7.3.4.2 Applications of HPTLC in cannabinoids analysis

Four cannabinoids, THC, CBD, CBN, and CBC, were identified and determined using HPTLC in two commercially available Japanese cannabis oils (hemp oil and Taima-Yu). The analysis was performed on RP-18 HPTLC plates using acetonitril 100% as a mobile phase. After development the plates were sprayed with a coloring agent (Echtbausalz B in 0.1 M NaOH). The limit of detection for the four cannabinoids was 50 µg/g [33].

An HPTLC-densitometry analytical method was developed for the screening and determination of the main neutral cannabinoids including Δ^9-THC, CBD, CBC, CBG, and THCV as well as quantification of Δ^9-THC and CBN in two decarboxylated medicinal cannabis cultivars. Si gel HPTLC plates (Merck, 20 × 10 cm silica gel 60 aluminum plates) were used, and the range of quantification was determined to be 50–500 ng at 206 nm. This method can be useful for forensic analysis, quality control of hemp, and quality control of medicinal Cannabis [34].

Chromatographic analysis of five neutral cannabinoids (Δ^9-THC, CBN, CBD, CBG, and CBC) has been carried out on amino HPTLC plates via over pressured-layer chromatographic technique on an OPLC BS 50 instrument. Thirty hemp samples were analyzed on a 10 cm × 20 cm plate within 4 min. Dichloromethane was used as a developing solvent and fast blue salt B as visualization reagent. Plates were evaluated by Desaga CD 60 slit scanner at a wavelength of 200 nm. To construct calibration curves, 0.10, 0.25, 0.50, 0.75, and 1.00 µg of each cannabinoid were applied to each plate [35].

Depending on the high-resolution capability of HPTLC, 10 different TLC mobile phase systems were used to figure out the most effective mobile phase for the analysis of cannabinoids in *C. sativa* products. Three major cannabinoids including Δ^9-THC, CBD, and CBN were recorded on all 10 systems. Reversed-phase RP-18 WF 254 10 × 10 cm plate was used for one solvent system, and normal-phase HPTLC silica gel 60 F254 20 × 10 cm plates were used for the rest nine systems. The optimum separation of the three major cannabinoids was achieved using xylene:hexane:diethylamine (25:10:1) and 6% diethylamine in toluene. This method could be applied to get better resolution for the three major cannabinoids in crime labs in spite of the complexity of the cannabis herbal products [36]. An HPTLC fingerprint method has been developed to differentiate between the cannabis varieties (high THC, intermediate, and high CBD) before and after decarboxylation by heating at 120 °C for 2 h. This method provides efficient resolution between 11 different cannabinoids. These cannabinoids include Δ^9-THC, CBD, CBN, CBC, CBG, THCA, CBGA, THCV, CBDV, CBDVA, and CBDA. Reversed-phase C18 HPTLC plates were used, and the bands were visualized by spraying with vanillin/H_2SO_4. Fast blue salt B, commonly used to visualize cannabinoids on normal phase plates, was not compatible with the C18 plates, particularly the Merck® plates. Silicycle C18 plates were more compatible with fast blue salt B visualizing reagent but did not provide as efficient separation of the different cannabinoids as did the Merck® plates. Stationary phases: Merck® C18 HPTLC plates (10 cm × 20 cm) were used. The plates were prewashed by developing in methanol and then dried at 120 °C for 30 min before use. Mobile phase used was: MeOH:H_2O: glacial acetic acid (80:10:10) [3]. Under the experimental conditions above, the

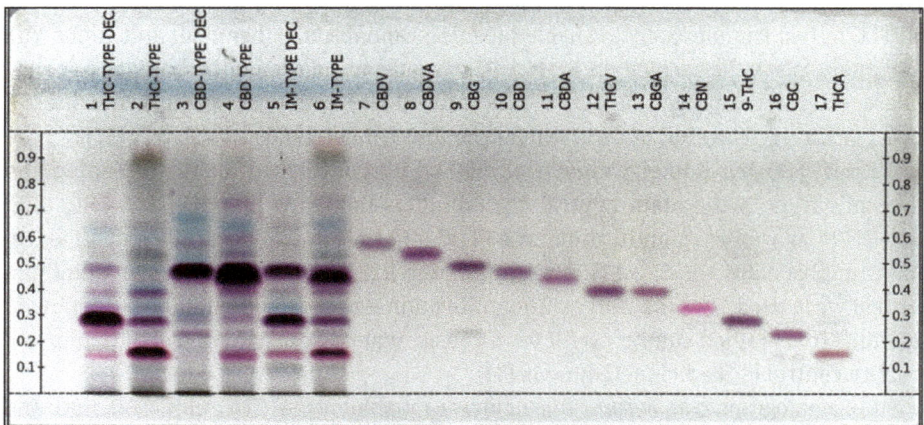

Figure 7.19: C18 HPTLC chromatogram of three cannabis varieties (before and after decarboxylation) and 11 cannabinoids (vanillin/H_2SO_4 spray reagent, white light).

different 11 cannabinoids were well separated as shown in Figure 7.19. The method can be used to distinguish between the three major varieties of cannabis (high THC/low CBD, intermediate variety with both THC and CBD at abundant levels and the CBD variety with high CBD/low THC levels) both in the native form and after decarboxylation.

7.4 Application of chromatographic analysis of cannabis for the potency monitoring program in the USA

7.4.1 Introduction

The potency monitoring program is a collaborative project between the United States DEA and the National Institute on Drug Abuse to monitor the potency (Δ^9-THC content) of confiscated marijuana, hashish, and hash oil to have a clear understanding of the composition of cannabis products available to the American public on the illicit market. Several other countries have similar programs.

Over the last few decades several reports have been published summarizing the data available for a given time frame [18, 37–41]. In this section, a summary of all the data available from January 1995 till December 31, 2021 is provided, giving a complete picture of the concentration of the major cannabinoids (Δ^9-THC and CBD) by year of seizure.

7.4.2 Samples

Samples analyzed in this report were confiscated during 1995 through 2021 by United States DEA. Samples provided by State law enforcement agencies are not included here since they do not fully represent materials on the illicit market. State samples might be seized prior to full maturation of the plants.

7.4.2.1 Sample identification

Sample classification is based on the physical characteristics according to the following guidelines:

Cannabis samples: All samples received are categorized as follows:
- *Marijuana* (known as herbal cannabis in Europe): usually found in four forms: (i) loose material – loose cannabis plant material with leaves, stems, and seeds; (ii) leaves – cannabis plant material consisting primarily of leaves;

(iii) kilo bricks – compressed cannabis with leaves, stems, and seeds (typical Mexican packaging); and (iv) buds – flowering tops of female plants with seeds.
- *Sinsemilla*: Flowering tops of unfertilized female plants with no seeds (subdivided as for marijuana with most samples being classified as buds).
- *Thai sticks*: Leafy material tied around a small stem (typical Thailand packaging).
- *Ditchweed*: Fiber-type wild cannabis found in the Midwestern region of the United States.

Hashish Samples – Hashish (known as cannabis resin in Europe) is composed of the resinous parts of the flowering tops of cannabis, mixed with some plant particles, and shaped into a variety of forms, for example, balls, sticks, or slabs. It is generally very hard with a dark green or brownish color.

Hash oil samples – Hash oil is a liquid or semisolid concentrated extract of cannabis plant material. Depending on the process used to prepare hash oil, it is usually dark green, amber, or brownish. Distilled extracts are golden yellow to light brown with much higher potency. Some refer to the distillate as shatter.

Sample storage

All samples are stored in a vault at controlled room temperature (17 ± 4 °C).

7.4.2.2 Samples preparation

Cannabis – Samples are manicured in a 14-mesh metal sieve to remove seeds and stems. Duplicate samples (2×0.1 g) are extracted with internal standard solution (ISTD) [3 mL, 4-androstene-3,17-dione (Sigma Aldrich, St. Louis, MO) as a (1 mg/mL) in chloroform/methanol (1:9, v/v) at room temperature for 1 h]. The extracts are transferred to GC vials via filtration through sterile cotton plugs, followed by capping of the vials.

Hashish – Samples are scrapped and powdered using a single-edge razor blade. Duplicate samples (2×0.1 g) are extracted following the procedure outlined for cannabis samples.

Hash Oil – Duplicate samples (2×50 mg) are extracted with 10 mL ISTD and sonicated for 10 min till the material dissolves. The extracts are then filtered using 5 ml syringe attached to 0.2 µm filter and then transferred to GC vials.

7.4.3 Analytical method

GC analyses are performed using Varian CP-3380 gas chromatographs, equipped with Varian CP-8400 autosamplers, capillary injectors, dual FIDs, and DB-1 MS columns

(15 m × 0.25 mm × 0.25 μm) (J&W Scientific, Folsom, CA). Data are recorded using a Dell Optiplex GX1 computer and Varian Star workstation software (version 6.1). Helium is used as carrier and detector makeup gas with an upstream indicating moisture trap and a downstream indicating oxygen trap. Hydrogen and compressed air are used as the combustion gases. The following instrument parameters are employed: air, 30 psi (300 mL/min); hydrogen, 30 psi (30 mL/min); column head pressure, 15 psi (1.0 mL/min); split flow rate, 100 mL/min; split ratio, 50:1; septum purge flow rate: 5 mL/min; makeup gas pressure, 20 psi (30 mL/min); injector temperature, 240 °C; detector temperature, 270 °C; oven program, 170 °C (hold 1 min) to 250 °C at 10 °C/min (hold 3 min); run time, 12 min; injection volume, 1 μL. The instruments are daily maintained and calibrated to ensure a Δ^9-THC/internal standard response factor ratio of 1.

Calculation of concentrations:

The concentrations of THC and CBD are calculated as follows:

$$\text{Cannabinoid \%} = \frac{\text{Area (cannabinoid)}}{\text{Area (ISTD)}} \times \frac{\text{amount (ISTD)}}{\text{amount (sample)}} \times 100$$

The method was validated and found to have linearity of 0.01 to 50% for THC and 0.025 to 50% for CBD with limit of quantitation of 0.01% for THC and 0.025% for CBD.

7.4.4 Results

Over the period of 1995–2021 a total of 42,966 cannabis samples confiscated by the DEA in the USA were analyzed for their cannabinoids content. A total of seven cannabinoids were analyzed, namely Δ^9-THC, CBD, CBG, CBC, CBN, THCV, and Δ^8-THC. However, only Δ^9-THC and CBD are included in this section in detail and their average concentration is reported by year of seizure (Table 7.1). Figure 7.20 shows the trend in the change of cannabis potency overtime, rising from ~4% in 1995 to 15.42% in 2021, almost a fourfold increase. In the same time frame, however, the CBD content tended to drop, especially in the last 10 years (Figure 7.21) (except 2020) resulting in a drastic change in the ratio of THC:CBD from about 14:1 in 1995 to 108:1 in 2017, and over 50:1 the years thereafter till 2021 (except 2020 where it was 21:1; Figure 7.22). This indicates that illicit cannabis growers are opting for high THC/low CBD varieties, especially for the recreational use.

Examining the % distribution of the number of samples for each year, with THC content of <3%, 3–7%, 7–12%, and e > 12% shows that the reason for the increase in the potency of cannabis overtime is the increase in the number of seizures with THC content of >12% (Figure 7.23). Considering the physical characteristics of the cannabis seized in the last few years, shows that the majority of seizures are made of sinsemilla which is the most potent cannabis botanical product. Not many seizures are made of the other types (buds, leaves, and loose material), with Thai sticks now disappearing

Table 7.1: Concentration of THC and CBD in DEA confiscated samples in the USA, 1995–2021 by year seized.

Year	n	THC (%)	CBD (%)	Ratio of THC and CBD
1995	3,763	3.96	0.28	14.1
1996	1,402	4.51	0.37	12.2
1997	1,337	5.01	0.41	12.2
1998	1,341	4.91	0.41	12.0
1999	1,825	4.60	0.42	11.0
2000	1,929	5.34	0.52	10.3
2001	1,687	6.11	0.55	11.1
2002	1,690	7.20	0.47	15.3
2003	1,872	7.15	0.47	15.2
2004	1,914	8.14	0.51	16.0
2005	2,295	8.02	0.49	16.4
2006	2,081	8.76	0.43	20.4
2007	2,143	9.58	0.46	20.8
2008	2,000	9.93	0.41	24.2
2009	2,073	9.75	0.39	25.0
2010	2,261	10.36	0.28	37.0
2011	2,342	11.13	0.22	50.6
2012	2,100	12.27	0.20	61.4
2013	1,230	11.98	0.16	74.9
2014	971	11.92	0.16	74.5
2015	1,003	10.91	0.18	60.6
2016	827	11.51	0.18	63.9
2017	777	14.08	0.13	108.3
2018	802	14.78	0.27	54.7
2019	448	13.67	0.21	65.1
2020	607	13.41	0.65	20.6
2021	245	15.42	0.29	53.2

from the market. Even kilobricks, traditionally made from entire mature flowering tops with leaves, seeds, and stems, are now made from sinsemilla.

The other cannabinoids of cannabis samples showed no particular pattern changes in concentration over the same period of time. For example, CBC levels did not change much over time (concentration ranging from 0.2 to 0.35%). On the other hand, since CBG is the first biosynthetic cannabinoid made in the plant which leads to all cannabinoids, including THC, it follows that its content in the plant material rises with the increased THC content. This is the case here, where the CBG levels show a trend to increase overtime and slowly rises from 0.13% in 1995 to ~0.5% in the last few years. As for CBN, the level of CBN reflects more the age of the plant material since it is actually a degradation product of THC by oxidation. Same with Δ^8-THC which is only found if any, in very low levels mostly less than 0.1%. THCV which is a naturally present cannabinoid with a three-carbon side chain is found in just about all samples albeit in very low levels averaging less than 0.1% on average over the period of this report.

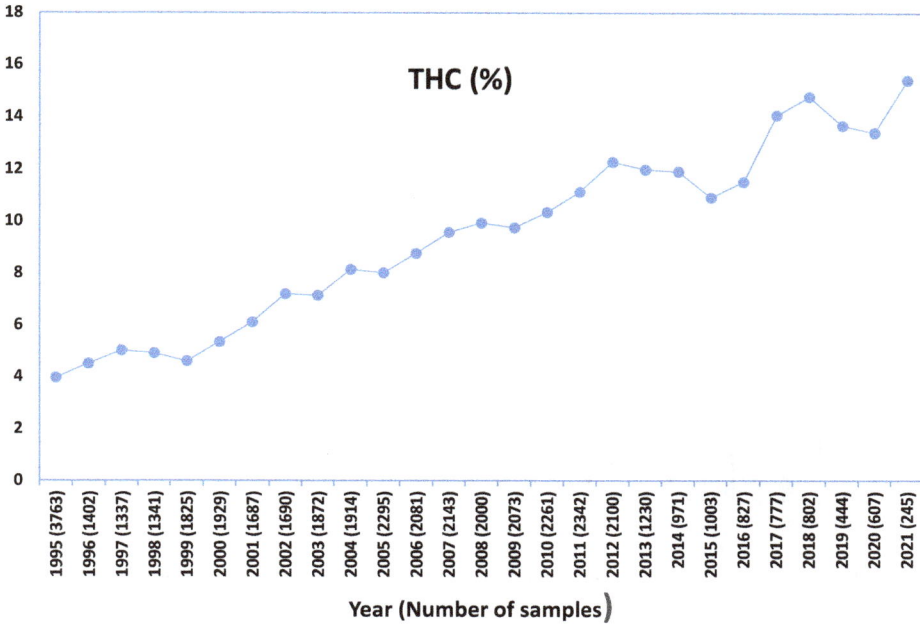

Figure 7.20: Average Δ⁹-THC concentration in DEA seized cannabis samples 1995–2021 by year seized.

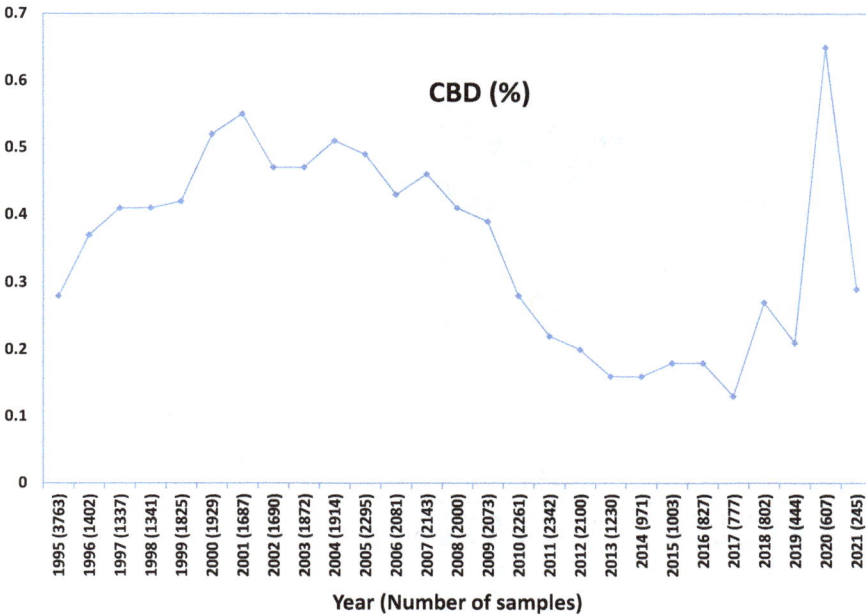

Figure 7.21: Average CBD concentration in DEA seized samples 1995–2021 by year seized.

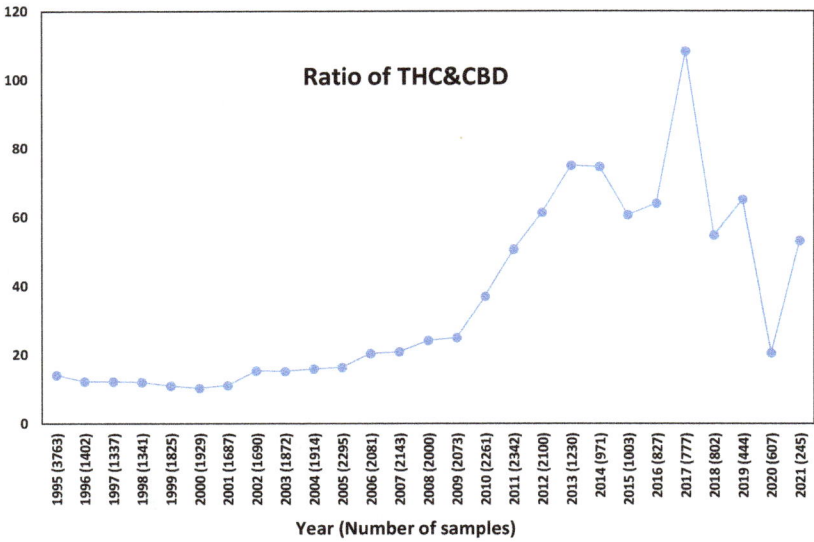

Figure 7.22: Change in the THC: CBD ratio in DEA seized cannabis samples 1995–2021 by year seized.

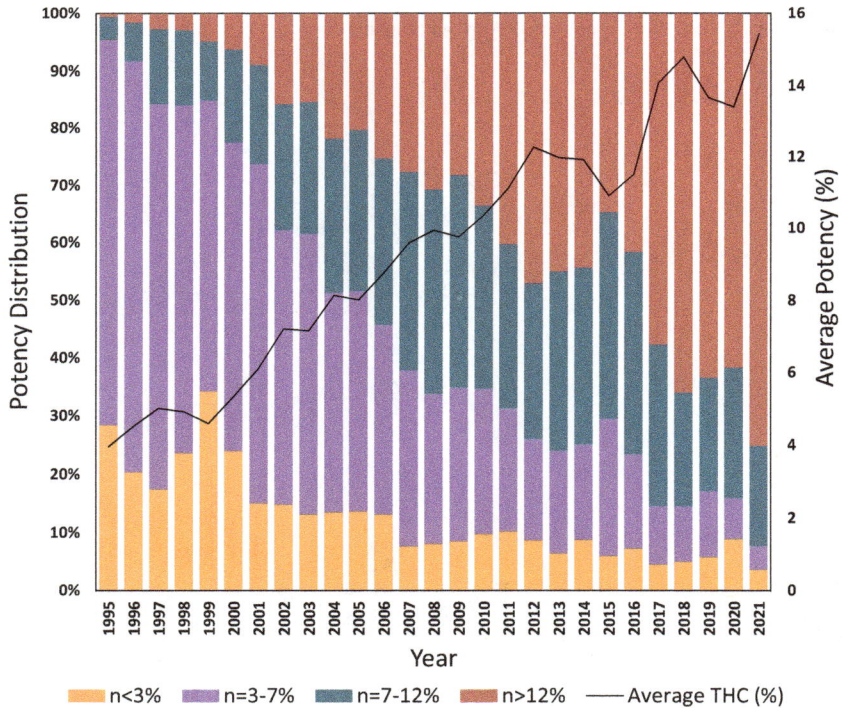

Figure 7.23: Potency distribution of DEA seized cannabis samples 1995–2021.

Other cannabis products include hashish and hash oil. Table 7.2 shows the cannabinoids content of hashish samples seized from 1995 to 2019 (with no seizures in 2020 or 2021). Traditionally hashish is manufactured from an intermediate variety (chemovar) of cannabis with balanced levels of THC and CBD, which is reflected in the final product. However, lately manufacturers have started manufacturing hashish from high THC chemovars, probably from leaves of the plant that are of relatively low potency than the flowers of the plant. This is reflected in Table 7.2 where early on the seized hashish had reasonable levels of CBD relative to THC while in the last 10 years the level of CBD dropped to mostly less than 1% while THC up to 40%.

Table 7.3 shows the cannabinoids content of hash oil. It appears that hash oil is mainly manufactured from high THC.

Table 7.2: Average hashish cannabinoid concentrations per year 1995 to 2019.

Year	n	Δ^8-THC (%)	Δ^9-THC (%)	CBD (%)	CBC (%)	CBN (%)	CBG (%)	THCV (%)
1995	19	0.00	3.60	3.34	0.51	1.70	0.32	0.14
1996	12	0.00	2.52	4.54	0.70	2.43	0.34	0.09
1997	31	0.00	8.92	3.99	0.75	2.13	0.48	0.32
1998	15	0.00	5.87	1.73	0.80	2.04	0.32	0.21
1999	23	0.00	4.94	1.79	0.59	2.07	0.51	0.30
2000	27	0.00	4.16	4.90	0.59	2.31	0.36	0.13
2001	13	0.00	8.48	2.72	0.58	1.54	0.58	0.30
2002	16	0.00	9.12	2.49	0.60	1.38	0.43	0.18
2003	16	0.00	9.23	3.92	0.68	1.85	0.41	0.16
2004	25	0.00	18.95	0.76	0.74	1.42	0.67	0.23
2005	47	0.00	11.98	1.67	0.86	1.91	0.37	0.17
2006	32	0.00	29.33	1.57	0.67	1.34	0.81	0.23
2007	70	0.00	27.71	1.19	0.80	1.84	0.97	0.30
2008	62	0.00	21.96	2.25	0.91	2.20	0.75	0.36
2009	42	0.06	17.40	1.47	0.92	2.74	0.40	0.15
2010	79	0.24	22.75	0.65	0.86	2.96	0.51	0.21
2011	120	0.24	30.04	0.51	1.12	2.97	0.70	0.23
2012	116	0.33	31.69	0.54	0.82	2.79	0.71	0.22
2013	41	0.28	29.28	0.36	0.72	2.44	0.83	0.21
2014	23	0.33	30.30	1.38	0.97	3.05	0.90	0.24
2015	23	0.15	17.62	0.95	0.51	2.60	0.56	0.14
2016	18	0.14	15.51	0.64	0.72	5.70	0.85	0.07
2017	10	0.30	40.89	0.27	0.82	2.83	1.13	0.24
2018	9	0.29	26.25	0.16	0.54	1.93	0.81	0.15
2019	1	0.53	37.88	0.10	0.53	0.84	1.80	0.26

Table 7.3: Average hash oil cannabinoid concentrations per year 1995 to 2021.

Year	N	Δ^8-THC (%)	Δ^9-THC (%)	CBD (%)	CBC (%)	CBN (%)	CBG (%)	THCV (%)
1995	13	0.00	13.23	0.72	1.00	4.24	0.51	0.34
1996	8	0.00	12.82	1.27	1.09	3.99	0.45	0.55
1997	10	0.00	18.20	0.28	1.05	3.51	0.35	0.57
1998	5	0.00	15.78	0.21	0.78	3.58	0.23	0.51
1999	11	0.00	16.21	0.38	1.34	4.83	0.33	0.35
2000	7	0.00	28.58	0.51	1.57	1.65	0.91	0.74
2001	7	0.00	19.44	1.32	1.20	4.36	0.95	0.58
2002	5	0.00	22.51	0.33	0.54	1.69	1.20	0.31
2003	4	0.00	15.54	0.19	0.80	1.32	0.33	0.39
2004	4	0.00	31.32	1.12	1.14	2.20	1.17	0.36
2005	6	0.00	6.40	0.35	0.22	1.15	0.15	0.18
2006	3	0.00	18.74	0.13	0.43	0.59	0.38	0.12
2007	18	0.00	24.85	0.57	0.89	1.53	0.65	0.31
2008	14	0.00	6.73	0.20	0.33	1.41	0.14	0.13
2009	6	0.08	12.71	0.34	0.21	3.28	0.26	0.24
2010	11	0.41	36.16	0.25	0.73	2.28	0.62	0.37
2011	22	0.21	35.28	0.42	0.91	2.41	0.82	0.30
2012	48	0.50	53.52	0.66	0.96	2.73	1.05	0.33
2013	32	0.37	49.98	0.46	1.06	2.18	1.05	0.27
2014	62	0.33	50.89	1.12	0.92	2.23	1.30	0.42
2015	55	0.53	56.26	0.57	1.14	2.84	1.60	0.29
2016	52	0.51	36.54	2.61	1.02	2.94	1.27	0.32
2017	42	0.59	55.18	0.59	1.02	2.62	1.53	0.35
2018	22	4.87	52.00	0.59	0.76	3.13	1.70	0.28
2019	55	4.35	53.43	3.09	0.68	2.96	1.70	0.31
2020	31	0.34	57.04	6.18	0.83	1.78	1.37	0.22
2021	25	4.09	51.38	5.18	0.62	2.00	1.47	0.46

References

[1] Joyce CRB, Curry SH. The botany and chemistry of Cannabis. London, UK, J. & A. Churchill, 1970.

[2] Small E, Marcus D. Hemp: A new crop with new uses for North America. Trends in New Crops and New Uses 2002; 24(5): 284–326.

[3] Sarma ND, et al. Cannabis inflorescence for medical purposes: USP considerations for quality attributes. Journal of Natural Products 2020; 83(4): 1334–1351.

[4] Pitt C, Hendron R, Hsia R. The specificity of the Duquenois color test for marihuana and hashish. Journal of Forensic Science 1972; 17(4): 693–700.

[5] Lau-Cam C, McDonnell J. The furfural test for cannabis: An evaluation and modification. Bulletin on Narcotics 1978; 30(2): 63–68.

[6] Kelly F, Krishna Addanki J, Bagasra O. The non-specificity of the Duquenois-Levine field test for marijuana. The Open Forensic Science Journal 2012; 5: 4–8.

[7] Thornton JI, Nakamura GR. The identification of marijuana. Journal of the Forensic Science Society 1972; 12(3): 461–519.

[8]   Fochtman FW, Winek CL. A note on the Duquenois-Levine test for marijuana. Clinical Toxicology
      1971; 4(2): 287–289.
[9]   Preedy VR. Handbook of cannabis and related pathologies: Biology, pharmacology, diagnosis, and
      treatment. Academic Pess, London, 2016.
[10] Korte F, Sieper H, Tira S. New results on hashish-specific constituents. Bulletin on Narcotics 1965; 17
      (1): 35–43.
[11] Kovar KA, Keilwagen S. Zur Kenntnis der Ghamrawy-Reaktion auf Haschisch und Marihuana. Archiv
      der Pharmazie 1984; 317(8): 724–732.
[12] Dos Santos NA, et al. Evaluating the selectivity of colorimetric test (Fast Blue BB salt) for the
      cannabinoids identification in marijuana street samples by UV–Vis, TLC, ESI (+) FT-ICR MS and ESI (+)
      MS/MS. Forensic Chemistry 2016; 1: 13–21.
[13] Hädener M, et al. Cannabinoid concentrations in confiscated cannabis samples and in whole blood
      and urine after smoking CBD-rich cannabis as a "tobacco substitute". International Journal of Legal
      Medicine 2019; 133(3): 821–832.
[14] Lewis K, et al. Validation of the 4-aminophenol color test for the differentiation of marijuana-type
      and hemp-type cannabis. Journal of Forensic Sciences 2021; 66(1): 285–294.
[15] Pellegrini M, et al. A rapid and simple procedure for the determination of cannabinoids in hemp
      food products by gas chromatography-mass spectrometry. Journal of Pharmaceutical & Biomedical
      Analysis 2005; 36(5): 939–946.
[16] Watanabe K. Cannabinoids and their metabolites. In: Drugs and poisons in humans. Edited by
      Osamu Suzuki and Kanako Watanabe. Springer, Berlin, Heidelberg. 2005; 187–194.
[17] de Oliveira GL, et al. Cannabinoid contents in cannabis products seized in São Paulo, Brazil,
      2006–2007. Forensic Toxicology 2008; 26(1): 31–35.
[18] ElSohly MA, et al. Changes in cannabis potency over the last 2 decades (1995–2014): Analysis of
      current data in the United States. Biological Psychiatry 2016; 79(7): 613–619.
[19] Ibrahim EA, et al. Determination of acid and neutral cannabinoids in extracts of different strains of
      Cannabis sativa using GC-FID. Planta Medica 2018; 84(04): 250–259.
[20] Ibrahim EA, et al. Analysis of terpenes in Cannabis sativa L. using GC/MS: Method development,
      validation, and application. Planta Medica 2019; 85(05): 431–438.
[21] Giese MW, et al. Method for the analysis of cannabinoids and terpenes in cannabis. Journal of AOAC
      International 2015; 98(6): 1503–1522.
[22] Gul W, et al. Determination of 11 cannabinoids in biomass and extracts of different varieties of
      Cannabis using high-performance liquid chromatography. Journal of AOAC International 2015; 98
      (6): 1523–1528.
[23] W. Xiaoyan, D. Mackowsky, J. Searfoss, M.J. Telepchak, Determination of cannabinoid content and
      pesticide residues in cannabis edibles and beverages, LC GC Eur 30 (2017) 16–21.
[24] United Nations Office on Drugs, a.C.. Recommended methods for the identification and analysis of
      Cannabis and Cannabis products: Manual for use by national drug testing laboratories. United
      Nations Publications, Vienna, 2009.
[25] De Backer B, et al. Innovative development and validation of an HPLC/DAD method for the
      qualitative and quantitative determination of major cannabinoids in cannabis plant material. Journal
      of Chromatography B 2009; 877(32): 4115–4124.
[26] Peschel W, Politi M. ¹H NMR and HPLC/DAD for Cannabis sativa L. chemotype distinction, extract
      profiling and specification. Talanta 2015; 140: 150–165.
[27] Izzo L, et al. Analysis of phenolic compounds in commercial Cannabis sativa L. inflorescences using
      UHPLC-Q-Orbitrap HRMS. Molecules 2020; 25(3): 631.
[28] Guillarme D, et al. Recent developments in liquid chromatography – Impact on qualitative and
      quantitative performance. Journal of Chromatography A 2007; 1149(1): 20–29.

[29] Heo S, et al. Simultaneous analysis of cannabinoid and synthetic cannabinoids in dietary supplements using UPLC with UV and UPLC–MS-MS. Journal of Analytical Toxicology 2016; 40(5): 350–359.

[30] Wang M, et al. Decarboxylation study of acidic cannabinoids: A novel approach using ultra-high-performance supercritical fluid chromatography/photodiode array-mass spectrometry. Cannabis and Cannabinoid Research 2016; 1(1): 262–271.

[31] Nahar L, Onder A, Sarker SD. A review on the recent advances in HPLC, UHPLC and UPLC analyses of naturally occurring cannabinoids (2010–2019). Phytochemical Analysis 2020; 31(4): 413–457.

[32] Badyal P, et al. Analytical techniques in simultaneous estimation: An overview. Austin Journal of Analytical and Pharmaceutical Chemistry 2015; 2(2): 1037.

[33] Yotoriyama M, et al. Identification and determination of cannabinoids in both commercially available and cannabis oils stored long term. Journal of Health Science 2005; 51(4): 483–487.

[34] Fischedick JT, et al. A qualitative and quantitative HPTLC densitometry method for the analysis of cannabinoids in *Cannabis sativa* L. Phytochemical Analysis 2009; 20(5): 421–426.

[35] Szabady B, Hidvegi E, Nyiredy S. Determination of neutral cannabinoids in hemp samples by overpressured-layer chromatography. Chromatographia 2002; 56(1): S165–S168.

[36] Liu Y, et al. High performance thin-layer chromatography (HPTLC) analysis of cannabinoids in cannabis extracts. Forensic Chemistry 2020; 19: 100249.

[37] ElSohly MA, et al. Constituents of *Cannabis sativa* L. XXIV: The potency of confiscated marijuana, hashish, and hash oil over a ten-year period. Journal of Forensic Science 1984; 29(2): 500–514.

[38] Mehmedic Z, et al. Potency trends of Δ^9-THC and other cannabinoids in confiscated cannabis preparations from 1993 to 2008. Journal of Forensic Sciences 2010; 55(5): 1209–1217.

[39] Chandra S, et al. New trends in cannabis potency in USA and Europe during the last decade (2008–2017). European Archives of Psychiatry and Clinical Neuroscience 2019; 269(1): 5–15.

[40] ElSohly MA, et al. A comprehensive review of cannabis potency in the United States in the last decade. Biological Psychiatry Cognitive Neuroscience and Neuroimaging 2021; 6(6): 603–606.

[41] ElSohly MA, et al. Potency trends of Δ 9-THC and other cannabinoids in confiscated marijuana from 1980–1997. Journal of Forensic Science 2000; 45(1): 24–30.

Index

https://doi.org/10.1515/9783110718362-008

www.ingramcontent.com/pod-product-compliance
Lightning Source LLC
Chambersburg PA
CBHW082110220326
41598CB00066BA/6283